Macmillan/McGraw-Hill • G

5

Math Triumphs

Book 1: Number and Operations and Algebra

Authors

Basich Whitney • Brown • Dawson • Gonsalves • Silbey • Vielhaber

Macmillan/McGraw-Hill
Glencoe

Photo Credits

Cover, i Carl Schneider/Taxi/Getty Images; **iv** (tl)File Photo, (tc tr)The McGraw-Hill Companies, (cl c)Doug Martin, (cr)Aaron Haupt, (bl bc)File Photo; **v** (L to R 1 2 3 4 6 7 8 9 11 12)The McGraw-Hill Companies, (5 10 13 14)File Photo; **vi** Larry Brownstein/Getty Images; **vi** Larry Brownstein/Getty Images; **vii** Comstock/PunchStock; **viii** Stockbyte/Getty Images; **1** Image Ideas/PictureQuest; **12** Pixtal/SuperStock; **26** Photodisc/Getty Images; **29** Jeffrey L. Rotman/Peter Arnold Inc.; **30** CORBIS; **38** Brooke Slezak/Taxi/Getty Images; **44** Lynn Betts/USDA Natural Resources Conservation Service; **45** Digital Archive Japan/Alamy; **51** Ken Davies/Masterfile; **53** PhotoLink/Getty Images; **54** C Squared Studios/Getty Images; **60** Gabe Palmer/CORBIS; **66** Stockdisc/PunchStock; **73** Purestock/AGE Fotostock; **74** Stockbyte/SuperStock; **76** Ken Cavanagh/The McGraw-Hill Companies; **80** Juniors Bildarchiv/Alamy; **81, 85** Ken Cavanagh/The McGraw-Hill Companies; **87** C Squared Studios/Getty Images; **90** LWA-Dann Tardif/CORBIS; **97** Matthew Scherf/iStockphoto; **98** Tony Cordoza/Alamy; **103** Burke/Triolo Productions/Getty Images; **104** Ryan McVay/Getty Images; **105** (t)Comstock/PunchStock, (b)Siede Preis/Getty Images; **106** Photodisc/Getty Images; **110** Brand X Pictures/PunchStock; **116** Rayman/Digital Vision/Getty Images; **131** Roy McMahon/CORBIS; **132** Dimitri Iundt/TempSport/CORBIS; **134** Siede Preis/Getty Images; **139** PhotoDisc/Getty Images; **142** F. Lukasseck/Masterfile; **150** CORBIS Super RF/Alamy; **157** Tony Freeman/PhotoEdit

The McGraw·Hill Companies

Macmillan/McGraw-Hill
Glencoe

Copyright © 2009 by The McGraw-Hill Companies, Inc. All rights reserved. Except as permitted under the United States Copyright Act, no part of this publication may be reproduced or distributed in any form or by any means, or stored in a database or retrieval system, without prior permission of the publisher.

Send all inquiries to:
Glencoe/McGraw-Hill
8787 Orion Place
Columbus, OH 43240-4027

ISBN: 978-0-07-888204-3
MHID: 0-07-888204-4

Math Triumphs
Grade 5, Book 1

Printed in the United States of America.

4 5 6 7 8 9 10 066 16 15 14 13 12 11 10 09

Math Triumphs

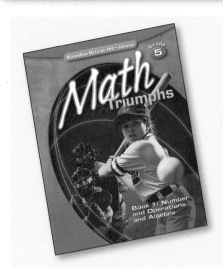

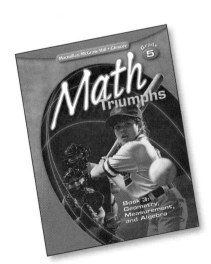

Authors and Consultants

AUTHORS

Frances Basich Whitney
Project Director, Mathematics K–12
Santa Cruz County Office of Education
Capitola, California

Kathleen M. Brown
Math Curriculum Staff Developer
Washington Middle School
Long Beach, California

Dixie Dawson
Math Curriculum Leader
Long Beach Unified
Long Beach, California

Philip Gonsalves
Mathematics Coordinator
Alameda County Office of Education
Hayward, California

Robyn Silbey
Math Specialist
Montgomery County Public Schools
Gaithersburg, Maryland

Kathy Vielhaber
Mathematics Consultant
St. Louis, Missouri

CONTRIBUTING AUTHORS

Viken Hovsepian
Professor of Mathematics
Rio Hondo College
Whittier, California

FOLDABLES Study Organizer | **Dinah Zike**
Educational Consultant,
Dinah-Might Activities, Inc.
San Antonio, Texas

CONSULTANTS

Assessment

Donna M. Kopenski, Ed.D.
Math Coordinator K–5
City Heights Educational Collaborative
San Diego, California

Instructional Planning and Support

Beatrice Luchin
Mathematics Consultant
League City, Texas

ELL Support and Vocabulary

ReLeah Cossett Lent
Author/Educational Consultant
Alford, Florida

Reviewers

Each person below reviewed at least two chapters of the Student Edition, providing feedback and suggestions for improving the effectiveness of the mathematics instruction.

Dana M. Addis
Teacher Leader
Dearborn Public Schools
Dearborn, MI

Renee M. Blanchard
Elementary Math Facilitator
Erie School District
Erie, PA

Jeanette Collins Cantrell
5th and 6th Grade Math Teacher
W.R. Castle Memorial Elementary
Wittensville, KY

Helen L. Cheek
K-5 Mathematics Specialist
Durham Public Schools
Durham, NCI

Mercy Cosper
1st Grade Teacher
Pershing Park Elementary
Killeen, TX

Bonnie H. Ennis
Mathematics Coordinator
Wicomico County Public Schools
Salisbury, MD

Sheila A. Evans
Instructional Support Teacher – Math
Glenmount Elementary/Middle School
Baltimore, MD

Lisa B. Golub
Curriculum Resource Teacher
Millennia Elementary
Orlando, FL

Donna Hagan
Program Specialist – Special Programs
    Department
Weatherford ISD
Weatherford, TX

Russell Hinson
Teacher
Belleview Elementary
Rock Hill, SC

Tania Shepherd Holbrook
Teacher
Central Elementary School
Paintsville, KY

Stephanie J. Howard
3rd Grade Teacher
Preston Smith Elementary
Lubbock, TX

Rhonda T. Inskeep
Math Support Teacher
Stevens Forest Elementary School
Columbia, MD

Albert Gregory Knights
Teacher/4th Grade/Math Lead Teacher
Cornelius Elementary
Houston, TX

Barbara Langley
Math/Science Coach
Poinciana Elementary School
Kissimmee, FL

David Ennis McBroom
Math/Science Facilitator
John Motley Morehead Elementary
Charlotte, NC

Jan Mercer, MA; NBCT
K-5 Math Lab Facilitator
Meadow Woods Elementary
Orlando, FL

Rosalind R. Mohamed
Instructional Support Teacher – Mathematics
Furley Elementary School
Baltimore, MD

Patricia Penafiel
Teacher
Phyllis Miller Elementary
Miami, FL

Lindsey R. Petlak
2nd Grade Instructor
Prairieview Elementary School
Hainesville, IL

Lana A. Prichard
District Math Resource Teacher K-8
Lawrence Co. School District
Louisa, KY

Stacy L. Riggle
3rd Grade Spanish Magnet Teacher
Phillips Elementary
Pittsburgh, PA

Wendy Scheleur
5th Grade Teacher
Piney Orchard Elementary
Odenton, MD

Stacey L. Shapiro
Teacher
Zilker Elementary
Austin, TX

Kim Wilkerson Smith
4th Grade Teacher
Casey Elementary School
Austin, TXL

Wyolonda M. Smith, NBCT
4th Grade Teacher
Pilot Elementary School
Greensboro, NC

Kristen M. Stone
3rd Grade Teacher
Tanglewood Elementary
Lumberton, NC

Jamie M. Williams
Math Specialist
New York Mills Union Free School District
New York Mills, NY

Contents

Chapter 1

Place Value and Number Relationships

Central Park, New York, New York

Chapter 2 **Multiplication**

Cypress Trees, Lake Bradford, Tallahassee, Florida

Contents

Chapter 3 — Division

Cape Fear, Wilmington, North Carolina

Chapter 4

Properties of Operations

Guadalupe Mountains, Salt Flat, Texas

Place Value and Number Relationships

Would you rather have $100,000 or $1,000,000?

To make that decision, you compare and order the numbers. You determine the value of the digits in each number.

STEP 1 Quiz

Math Online Are you ready for Chapter 1? Take the Online Readiness Quiz at *glencoe.com* to find out.

STEP 2 Preview

Get ready for Chapter 1. Review these skills and compare them with what you'll learn in this chapter.

What You Know	What You Will Learn
Suppose someone gave you $1,025. You could count it to determine how much you had.	*Lesson 1-1* You can write $1,025 as: **One thousand twenty-five dollars** or **1,000 + 20 + 5 dollars**
You know that $250 is more than $205. So, Bike 1 costs more than Bike 2.  Bike 1  Bike 2	*Lesson 1-1* The placement of the 5 in the numbers makes a big difference. $2<u>5</u>0 →  $20<u>5</u> → 
You know that if you have 5 dimes, you can count them in this way: 10¢　20¢　30¢　40¢　50¢	*Lessons 1-3, 1-4* A **number pattern** is a regular, repeating sequence of numbers. **10, 20, 30, 40, 50** The pattern is to add 10 to each number.

Whole Numbers Less Than 10,000

KEY Concept

Place value tells you the value of each digit in a number. The number 1,256 is in **standard form**.

1,256

	thousands	hundreds	tens	ones
Standard Form	1	2	5	6
Model				
Value	$1 \times 1{,}000 = 1{,}000$	$2 \times 100 = 200$	$5 \times 10 = 50$	$6 \times 1 = 6$

In **expanded form** the number 1,256 is

$1{,}000 + 200 + 50 + 6$

VOCABULARY

expanded form
the form of a number as a sum that shows the value of each digit

place value
the value given to a digit by its position in a number

standard form
writing a number using only digits

The **expanded form** of a number shows the value of each digit. It uses the operation of addition.

Example 1

Identify the value of the underlined digit in 3,<u>8</u>15.

1. Write the number in a place-value chart.

2. The underlined digit is in the hundreds place.

3. Multiply the underlined digit by the value of its place.
   $8 \times 100 = 800$

4. The underlined digit has a value of 800.

1000	100	10	1
thousands	hundreds	tens	ones
3	<u>8</u>	1	5

YOUR TURN!

Identify the value of the underlined digit in 6,5_2_5.

1. Write the number in a place-value chart.

2. In what place is the underlined digit? _____

3. Multiply the underlined digit by the value of its place.

   _____ × _____ = _____

4. What is the value of the underlined digit? _____

1000	100	10	1
thousands	hundreds	tens	ones

Example 2

Write 7,054 in expanded form.

1. Write the number in a place-value chart.

1000	100	10	1
thousands	hundreds	tens	ones
7	0	5	4
7 × 1,000	0 × 100	5 × 10	4 × 1
7,000	0	50	4

> Do not include a value of 0 in expanded form.

2. Multiply each digit by the value of its place.

   7 × 1,000 = 7,000
   5 × 10 = 50
   4 × 1 = 4

3. Write the values as an addition expression.
   7,000 + 50 + 4

YOUR TURN!

Write 9,607 in expanded form.

1. Write the number in a place-value chart.

1000	100	10	1
thousands	hundreds	tens	ones

2. Multiply each digit by the value of its place.

3. Write the values as an addition expression.

   _____

Who is Correct?

What is the value of the underlined digit in _5_,085?

LaBron
5,000

Adita
500

Jolene
50

Circle correct answer(s). Cross out incorrect answer(s).

GO ON

▶ Guided Practice

Identify the digit in the thousands place-value position of each number.

1 4,908 _____

2 2,005 _____

Step by Step Practice

3 Identify the value of the underlined digit in 5,470.

Step 1 Write the number in a place-value chart.

Step 2 The place value of the 5 is _____.

Step 3 Multiply 5 × _____.

Step 4 The underlined digit has a value of _____.

1000	100	10	1
thousands	hundreds	tens	ones

Identify the value of each underlined digit.

4 2,0<u>8</u>7

1000	100	10	1
thousands	hundreds	tens	ones

_____

5 8,<u>3</u>62 _____

6 3,<u>4</u>91 _____

7 9<u>8</u>4 _____

8 3<u>2</u>2 _____

9 <u>9</u>,873 _____

10 <u>5</u>,060 _____

Write each number in expanded form.

11 6,915

_____

12 3,654

_____

13 5,107

_____

14 3,806

_____

15 6,300

_____

16 2,500

_____

Write each number in standard form.

17 1,000 + 700 + 50 + 2 _____

18 4,000 + 700 + 10 + 7 _____

19 3,000 + 6 _____

20 5,000 + 5 _____

Step by Step Problem-Solving Practice

21 **BASEBALL** An umpire picks four players with single-digit uniform numbers from each team. The team with the greatest number value will bat first. The Ravens' numbers are 2, 6, 9, and 5. The Hawks' numbers are 1, 8, 5, and 7. Which team will bat first?

Understand Read the problem. Write what you know.

The numbers for the Ravens are _____, _____, _____, and _____.

The numbers for the Hawks are _____, _____, _____, and _____.

Plan Pick a strategy. One strategy is to make a table. Use place-value charts to make each team's greatest number.

1000	100	10	1
thousands	hundreds	tens	ones

Ravens

Solve Pick a number from each team that will give the highest value in the thousands place.

Ravens _____ Hawks _____

Pick a number from the remaining digits for each team that will give the highest value in the hundreds, tens, and ones places.

The Ravens' greatest number is _____.

The Hawks' greatest number is _____.

The _____ will bat first.

1000	100	10	1
thousands	hundreds	tens	ones

Hawks

Check Rearrange each team's number using a new thousands-place digit. Are the new numbers greater than the original numbers?

22 **COMPUTERS** Corinna has to change her current computer password. In order to remember it, she wants to keep the same digits, but lower the value of the numbers altogether. Her current password is 3762. What will be her new password? Check off each step.

_____ Understand: I circled key words.

_____ Plan: To solve the problem, I will _____.

_____ Solve: Her new password will be _____.

_____ Check: I checked my answer by _____.

GO ON →

23 **Reflect** Write a four-digit number. Identify the place of each digit and its value.

_____

_____

_____

 **Skills, Concepts, and Problem Solving**

Identify the value of the underlined digit.

24 4,78<u>6</u> _____

25 8,<u>3</u>33 _____

26 2,1<u>4</u>7 _____

27 <u>5</u>,740 _____

Write each number in expanded form.

28 4,312 _____

29 6,753 _____

30 7,007 _____

31 3,100 _____

Write each number in standard form.

32 4,000 + 500 + 6 _____

33 6,000 + 700 + 70 + 7 _____

34 9,000 + 20 _____

35 5,000 + 400 + 4 _____

Solve.

36 **PUZZLES** Use the digits 7, 9, and 8 to write the greatest possible even number. Use each digit only once. _____

Vocabulary Check **Write the vocabulary word that completes each sentence.**

37 The operation of _____ is used in expanded form.

38 The _____ place is to the left of the hundreds place.

39 **Writing in Math** Create a place-value chart for 9,762. Be sure to put in the titles for the place values used. Explain why you placed each digit in its location on the chart.

_____

_____

STOP

Read and Write Whole Numbers in the Millions

KEY Concept

Place values are grouped into **periods**.

2,436,819

millions			thousands			ones		
1,000,000	100,000	10,000	1,000	100	10	1
millions	hundred thousands	ten thousands	thousands	hundreds	tens	ones
2	4	3	6	8	1	9

Say: two million, four hundred thirty-six thousand, eight hundred nineteen.

VOCABULARY

period
   a group of three digits in the place-value chart

standard form
   writing a number using only digits

word form
   a way to write numbers using only words

Periods are separated by commas. The word "one" is not said at the end of the ones period.

Example 1

Write two million, six hundred five thousand, one hundred seventy-three in standard form.

1. Rewrite the number using digits and periods.

   2 million, 605 thousand, 173

2. Fill in each period of a place-value chart. Write the number. 2,605,173

millions			thousands			ones		
1,000,000	100,000	10,000	1,000	100	10	1
millions	hundred thousands	ten thousands	thousands	hundreds	tens	ones
2	6	0	5	1	7	3

YOUR TURN!

Write seven million, five hundred seventy-five thousand, two in standard form.

1. Rewrite the number using digits and periods.

   _____

2. Fill in each period of a place-value chart.

   Write the number. _____

millions			thousands			ones		
1,000,000	100,000	10,000	1,000	100	10	1
millions	hundred thousands	ten thousands	thousands	hundreds	tens	ones

GO ON

Example 2

Write 4,560,326 in word form.

1. Rewrite the number using digits and periods.
   4 million, 560 thousand, 326

2. Write the words for the millions period.
   four million

3. Write the words for the thousands period.
   five hundred sixty thousand

4. Write the words for the ones period.
   three hundred twenty-six

5. Write the periods in order. Separate each period with a comma.
   four million, five hundred sixty thousand, three hundred twenty-six

YOUR TURN!

Write 9,592,030 in word form.

1. Rewrite the number using digits and periods. _____

2. Write the words for the millions period. _____

3. Write the words for the thousands period. _____

4. Write the words for the ones period. _____

5. Write the periods in order. Separate each period with a comma.

   _____

Who is Correct?

Write 2,200,002 in word form.

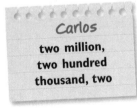

Carlos

two million,
two hundred
thousand, two

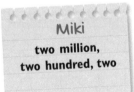

Miki

two million,
two hundred, two

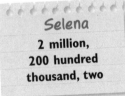

Selena

2 million,
200 hundred
thousand, two

Circle correct answer(s). Cross out incorrect answer(s).

▶ Guided Practice

Use the place-value chart to answer each question. Then write each number in the chart.

1,000,000	100,000	10,000	1,000	100	10	1
millions	hundred thousands	ten thousands	thousands	hundreds	tens	ones
1.						
2.						
3.						

1 How many zeros are in 5 millions? ____

2 How many zeros are in 7 hundred thousands? ____

3 How many zeros are in 3 ten thousands? ____

Write the missing number in each equation.

4 $1,000,000 + 200,000 + 6,000 + 70 +$ _____ $= 1,206,079$

5 $2,000,000 +$ _____ $+ 40,000 + 800 + 1 = 2,340,801$

Write each number in standard form.

6 three million, four hundred forty thousand, five hundred ten _____

7 seven hundred thousand, seven hundred _____

Step by Step Practice

8 Write 305,660 in word form.

> When the digit is 0, its value is zero, so you do not write that place value.

Step 1 Rewrite the number using digits and periods.

_____ thousand, _____

Step 2 Write the words for the thousands period. _____

Step 3 Write the words for the ones period. _____

Step 4 Write the periods together. _____

Write each number in word form.

9 4,203,915 _____ million _____ thousand _____

_____

10 50,500,005

_____

GO ON ▶

Solve.

11 **PHYSICS** The speed of light is one hundred eighty-six thousand, two hundred eighty-two miles per second. What is this number in standard form?

Understand	Read the problem. Write what you know. What is the greatest period in the speed of light? _____
Plan	Pick a strategy. One strategy is to solve a simpler problem.
Solve	Break the number into the thousands period and the ones period.
	Write the thousands period in digits. _____
	Write the ones period in digits. _____
	Write the periods together, separated by a comma. _____
Check	Read the standard form aloud. Follow along with the word form to make sure the forms match.

12 **SPACE** The distance between two asteroids is three hundred eighty-four thousand, three hundred eighty-five kilometers. Write this distance in standard form. Check off each step.

_____ Understand: I circled key words.

_____ Plan: To solve the problem, I will _____.

_____ Solve: The answer is _____.

_____ Check: I checked my answer by _____.

13 **GEOGRAPHY** Look at the photo at the right. Write the height of Mount Everest in word form.

_____

29,035 feet

Mount Everest

14 **EARTH SCIENCE** The Mariana Trench in the Pacific Ocean is the deepest part at 36,198 feet below sea level. Write this number in word form.

_____

_____

15 **Reflect** Place commas in the number 3710392. Then write the number in word form.

_____

 ## Skills, Concepts, and Problem Solving

Use the place-value chart to answer each question. Then write each number in the chart.

16 How many zeros are in 7 ten thousands?

_____

17 How many zeros are in 2 millions?

_____

18 How many zeros are in 4 hundred thousands?

_____

1,000,000	100,000	10,000	1,000	100	10	1
millions	hundred thousands	ten thousands	thousands	hundreds	tens	ones
16.						
17.						
18.						

Write the missing number in each equation.

19 $2{,}000{,}000 + 400{,}000 + 30{,}000 + 6{,}000 + 800 + 10 +$ _____ $= 2{,}436{,}814$

1,000,000	100,000	10,000	1,000	100	10	1
millions	hundred thousands	ten thousands	thousands	hundreds	tens	ones
2	4	3	6	8	1	?

20 _____ $+ 60{,}000 + 500 + 5 = 1{,}060{,}505$

Write each number in standard form.

21 three hundred eight thousand, six hundred fifty-three _____

22 six hundred six thousand, four hundred sixty-five _____

23 one hundred one thousand, one hundred one _____

24 five hundred five thousand, five hundred _____

Write each number in word form.

25 1,305,450

_____

26 3,407,690

_____

27 8,211,099

_____

Solve.

28 GEOGRAPHY   Greenland is the largest island in the world.
Its area is eight hundred forty thousand square miles.
Write this number in standard form.

_____

Area = one hundred forty-one thousand, two hundred five square kilometers

29 GEOGRAPHY   Write the area of the state of New York
in standard form, using the map at the right.

_____

New York

**Vocabulary Check   Write the vocabulary word that
completes each sentence.**

30 Commas are used to separate _____ in a number.

31 A number that has a 1 followed by six 0s is called one _____.

32 Writing in Math   Divide a sheet of paper into four sections. Write
the number 5,345,600 in one section of the paper. Use the other
three sections to represent the number in three other ways.

_____

_____

▶ **Spiral Review**

Identify the place-value position of each underlined digit. (Lesson 1-1, p. 4)

33 2,7̲02 _____

34 3̲00,594 _____

35 203̲,000 _____

36 9,26̲0 _____

STOP

14   Chapter 1 Place Value and Number Relationships

Copyright © by The McGraw-Hill Companies, Inc.

Progress Check 1 (Lessons 1-1 and 1-2)

Write each number in expanded form.

1 1,802 _____

2 3,461 _____

3 7,015 _____

4 9,060 _____

5 2,544 _____

6 7,178 _____

7 4,881 _____

8 6,836 _____

Write each number in word form.

9 8,324,015 _____

10 1,447,398 _____

11 6,700,000 _____

Write each number in standard form.

12 three million, four hundred five thousand _____

13 five million, one hundred twelve thousand _____

Solve.

14 **PUZZLES**   Use the digits 1, 2, and 4 to write the least possible odd number. Use each digit once.

_____

15 **PUZZLES**   Use the digits 6, 8, and 5 to write the least possible odd number. Use each digit once.

_____

16 **POPULATION**   The population of Chicago in 2010 is estimated to become two million, eight-hundred thousand. Write this number in standard form.

_____

Number Relationships

KEY Concept

Number **patterns** follow a rule.

You can use a **rule** to answer questions about the pattern and to predict what comes next.

12, 24, 36, 48, 60, 72, 84, … The rule is "Add 12."

To continue the pattern, add 12 to the last term.

$$84 + 12 = 96$$

$$96 + 12 = 108$$

$$108 + 12 = 120$$

So, the next three terms in the pattern are 96, 108, and 120.

Rules define relationships between numbers. For example, there are 12 inches in 1 foot. So, there are 24 inches in 2 feet (12×2), 36 inches in 3 feet (12×3), 48 inches in 4 feet (12×4), and so on.

VOCABULARY

pattern
    a sequence of numbers, figures, or symbols that follows a rule or design

rule
    tells how numbers are related to each other

Sometimes patterns can follow more than one rule.

Example 1

A car has 4 wheels. How many wheels are on 6 cars?

1. Each car has 4 wheels.

2. One rule is "add 4 for each car."

    $4 + 4 + 4 + 4 + 4 + 4 = 24$

3. Another rule is "multiply the number of cars by 4."

    $6 \times 4 = 24$

4. No matter which rule is used, there are 24 wheels on 6 cars.

YOUR TURN!

A spider has 8 legs. How many legs are on 5 spiders?

1. Each spider has _____ legs.

2. One rule is _____.

    _____ + _____ + _____ + _____ + _____ = _____.

3. Another rule is _____
    _____.

    _____ × 8 = _____

4. There are _____ legs on 5 spiders.

Example 2

Write the next three terms in the pattern.

1, 3, 7, 15

1. Find the rule.
   The rule is multiply by 2, and then add 1.
   $1 \times 2 = 2 + 1 = 3$
   $3 \times 2 = 6 + 1 = 7$
   $7 \times 2 = 14 + 1 = 15$

2. Continue the pattern.
   $15 \times 2 = 30 + 1 = 31$
   $31 \times 2 = 62 + 1 = 63$
   $63 \times 2 = 126 + 1 = 127$

The next three terms are 31, 63, 127.

YOUR TURN!

Write the next three terms in the pattern.

4, 7, 13, 25

1. Find the rule.
   The rule is _____
   _____
   _____
   _____

2. Continue the pattern.
   _____
   _____
   _____

The next three terms are _____, _____,
and _____.

Who is Correct?

How many cups are in 5 pints?

Number of Pints	1	2	3	4	5
Number of Cups	2	4	6	8	

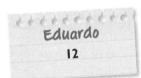

Eduardo

12

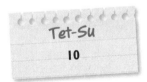

Tet-Su

10

Patrick

100

Circle correct answer(s). Cross out incorrect answer(s).

 Guided Practice

Find a rule for each pattern.

1 20, 15, 10, 5 _____

2 10, 17, 24, 31 _____

3 200, 100, 50, 25 _____

4 3, 9, 27, 81 _____

GO ON

5 There are 36 inches in 1 yard and 72 inches in 2 yards. Continue the pattern to find how many inches are in 3, 4, and 5 yards.

Step 1   Write the pattern with the missing terms

36, 72, _____, _____, _____

Step 2   Find a rule: Multiply by _____.

Step 3   Multiply the number of yards by _____ to continue the pattern.

$3 \times$ _____ = _____ $4 \times$ _____ = _____

$5 \times$ _____ = _____

The next three terms are _____, _____, and _____.

Find a rule in each sequence. Then write the next three terms.

6 19, 16, 13, 10, _____, _____, _____

Rule: _____

$10 -$ _____ = _____

_____ $-$ _____ = _____

_____ $-$ _____ = _____

The next three terms are:

_____, _____, and _____.

7 320, 160, 80, 40, _____, _____, _____

Rule: _____

$40 \div$ _____ = _____

_____ $\div$ _____ = _____

_____ $\div$ _____ = _____

The next three terms are:

_____, _____, and _____.

8 1, 3, 9, 27, _____, _____, _____

Rule: _____

The next three terms are:

_____, _____, and _____.

9 153, 162, 171, 180, _____, _____, _____

Rule: _____

The next three terms are:

_____, _____, and _____.

Write the next three conversions in each pattern.

10.

Number of Pounds	1	2	3	4
Number of Ounces	16			

11.

Number of Feet	1	2	3	4
Number of Inches	12			

Step by Step Problem-Solving Practice

Solve.

Problem-Solving Strategies
☑ Make a table.
☐ Guess and check.
☐ Act it out.
☐ Solve a simpler problem.
☐ Work backward.

12. **BOOKS** Fidel bought 6 books. The first book cost $5. Each additional book cost $1 more than the previous book. How much did Fidel spend on books in all?

Understand   Read the problem. Write what you know.
The first book costs _____.
Each additional book costs _____ more than the previous book.
Fidel bought _____ books.

Plan   Pick a strategy. One strategy is to make a table. Label the rows Book and Cost.

Solve   One book costs _____. The cost increases by _____ for each additional book.
The rule is _____.

Book	1	2	3	4	5	6
Cost	$	$	$	$	$	$

To find the total cost, add the cost of each book.

_____ + _____ + _____ + _____ + _____ + _____ = _____
Book 1   Book 2   Book 3   Book 4   Book 5   Book 6   Total

Fidel spent _____ on books.

Check   Does your answer make sense?

GO ON

13 FISH   A swordfish grows at a regular rate for the first year of life. Suppose it weighs 14 pounds at the age of 1 month, 28 pounds at the age of 2 months, and 42 pounds at the age of 3 months. What is the weight of a swordfish at the age of 6 months? Check off each step.

_____ Understand: I circled key words.

_____ Plan: To solve the problem, I will _____.

_____ Solve: The answer is _____.

_____ Check: I checked my answer by _____.

14 FITNESS   Martha runs for 30 minutes each day except for Saturday and Sunday. After 2 weeks, how much time will Martha have spent running?

_____

15 Reflect   Explain a rule for the pattern 10, 20, 40, 80.

_____

_____

_____

 Skills, Concepts, and Problem Solving

Find a rule for each pattern.

16 625, 125, 25, 5 _____

17 561, 574, 587, 600 _____

18 2, 6, 18, 54 _____

19 482, 346, 210, 74 _____

20 9, 36, 144, 576 _____

21 1080, 180, 30, 5 _____

In each sequence, find a rule. Then write the next three terms.

22 18, 30, 42, 54

Rule: _____

Next terms: _____, _____, _____

23 321, 310, 299, 288

Rule: _____

Next terms: _____, _____, _____

24 7, 14, 28, 56

Rule: _____

Next terms: _____, _____, _____

25 31,250; 6,250; 1,250; 250

Rule: _____

Next terms: _____, _____, _____

26 101, 110, 119, 128

Rule: _____

Next terms: _____, _____, _____

27 5103, 1701, 567, 189

Rule: _____

Next terms: _____, _____, _____

28 Write the next three conversions in the pattern.

Number of Gallons	1	2	3	4
Number of Quarts	4			

29 Write the next three conversions in the pattern.

Number of Meters	1	2	3	4
Number of Centimeters	100			

Solve.

30 **BOOKS** Dimitri placed boxes of books on 5 shelves in the library. He put 1 box on the top shelf, 3 boxes on the second shelf, and 5 boxes on the third shelf. If he continues this pattern, how many boxes will he put on the fifth shelf? _____

31 **MUSIC** For a band concert, chairs were set up for the musicians. There were 4 chairs in every row. How many chairs are there in 5 rows? _____

1st row	2nd row	3rd row	4th row	5th row
4 chairs	8 chairs			

Copyright © by The McGraw-Hill Companies, Inc.

Vocabulary Check   **Write the vocabulary word that completes each sentence.**

32   A(n) _____ tells how numbers are related to each other.

33   A(n) _____ is a sequence of numbers, figures, or symbols that follows a rule or design.

34   Writing in Math   Explain how to find the next three terms in the sequence 2, 4, 8, 16 using two different rules.

_____

_____

▶ Spiral Review

Identify the value of the underlined digit. (Lesson 1-1, p. 4)

35   6,3<u>9</u>1 _____

36   <u>8</u>,404 _____

37   9,00<u>8</u> _____

38   1,<u>9</u>82 _____

Write each number in standard form. (Lesson 1-2, p. 9)

39   nine hundred two thousand, three hundred sixty-one _____

40   four hundred thirty-two thousand, one hundred five _____

Write each number in standard form. (Lesson 1-2, p. 9)

41   2,492,102 _____

42   1,973,923 _____

_____

Solve. (Lesson 1-2, p. 9)

43   SCHOOL   Daniel wrote 5,305,707 in words on his homework paper. He wrote "five million, three hundred five thousand, seven hundred." What mistake did Daniel make?

_____

_____

STOP

Linear Patterns

KEY Concept

Patterns follow a **rule**. A rule describes the relationship that one element of a sequence has with the next element of the sequence. A rule can also describe the relationship an element has with its position in the sequence.

VOCABULARY

pattern
 a sequence of numbers, figures, or symbols that follows a rule or design

rule
 tells how numbers, figures, or symbols in a pattern are related to each other

A table can help you identify and extend a pattern by its rules.

+1 +1 +1 +1 +1

Number of Ducks	1	2	3	4	5	6
Number of Feet	2	4	6	8	10	12

+2 +2 +2 +2 +2

For each additional duck, the number of feet increases by 2.

Another way to describe the pattern is the number of ducks multiplied by 2 is the number of feet.

GO ON

Example 1

Margo rides her bike at 6 miles per hour. How many miles can she travel in 4 hours?

1. Make a table.

2. For every 1 hour traveled, Margo rides 6 miles. So the number of miles traveled increases by 6 for each additional hour. The rule is add 6 for each hour traveled.

3. Add 6 to a term to obtain the next term in the pattern.

+1   +1   +1

Number of Hours	1	2	3	4
Number of Miles Traveled	6	12	18	24

+6   +6   +6

Margo can ride 24 miles in 4 hours.

YOUR TURN!

Dawn's Diner serves 9 cheesesticks in 1 order. How many cheesesticks are in 6 orders?

1. Make a table.

2. Each order has _____ cheesesticks. The number of cheesesticks increases by _____ for each additional order. The rule is _____ for each order.

3. _____ to a term to obtain the next term in the pattern.

+1 +1 +1 +1 +1

Number of Orders	1	2	3	4	5	6
Number of Cheesesticks						

+9 +9 +9 +9 +9

There are _____ cheesesticks in 6 orders.

Example 2

For each block, there are 5 apartment buildings. How many apartment buildings are in 8 blocks?

1. For each 1 block, there are 5 apartment buildings. The number of buildings increases by 5 for each additional block. The rule is multiply by 5.

2. Multiply the number of blocks by the number of apartment buildings in each block.

$8 \times 5 = 40$

There are 40 apartment buildings in 8 blocks.

YOUR TURN!

For each car, there are 4 wheels. How many wheels do 10 cars have?

1. For each 1 car, there are _____ wheels. The number of wheels increases by _____ for each additional car. The rule is _____.

2. Multiply the number of cars by the number of wheels on each car.

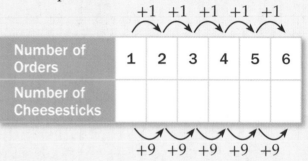
number   number
of cars   of wheels

There are _____ wheels on 10 cars.

Who is Correct?

How many fingers (including the thumb) do 3 people have?

Gilda
$10 \times 1 = 10$

Ian
$3 \times 8 = 24$

Marni
$10 \times 3 = 30$

Circle correct answer(s). Cross out incorrect answer(s).

▶ Guided Practice

Write a possible situation for each rule.

1 Add 2 for each additional person.

_____

2 Multiply by 4 for each additional animal.

_____

3 **EXERCISE** John jogs 2 miles in 20 minutes. How long will it take John to jog 8 miles?

The rule is add _____ for every _____ miles.

It will take John _____ minutes to jog 8 miles.

		+2	+2	+2
Number of Miles	2	4	6	8
Number of Minutes	20	40		

Step by Step Practice

4 There are 3 feet in a yard. How many feet are there in 5 yards?

Step 1 Each yard has _____ feet. The number of feet increases by _____ for each additional yard. The rule is _____.

Step 2 Multiply the number of yards by the number of feet in each yard.

_____ × _____ = _____

There are _____ feet in 5 yards.

GO ON

5 READING   Francisco can read at a rate of 30 pages for every 2 hours. How many pages can Francisco read in 4 hours?

Francisco can read _____ pages in 4 hours.

6 SWIMMING   Liz swam at a rate of 18 laps per hour. How many laps will Liz swim in 3 hours?

Liz will swim _____ laps in 3 hours.

Step by Step Problem-Solving Practice

Solve.

Problem-Solving Strategies
☑ Use logical reasoning.
☐ Guess and check.
☐ Act it out.
☐ Solve a simpler problem.
☐ Work backward.

7 FUNDRAISING   Larisa sold raffle tickets for a fundraiser. She sold the tickets at a rate of 6 tickets each hour. How many tickets did Larisa sell in 5 hours?

Understand   Read the problem. Write what you know.

Larisa sold _____ tickets each hour.

Plan   Pick a strategy. One strategy is to use logical reasoning.
Multiply the number of tickets Larisa sold each hour by the number of hours.

Solve   Multiply.

_____ × _____ = _____
number   number   total tickets
of tickets   of hours

Larisa sold _____ tickets in 5 hours.

Check   Skip count by 6 five times.

_____ + _____ + _____ + _____ + _____ = _____

8 NATURE   How many minutes will it take for Lupé to climb out of a hole that is 27 inches deep? Check off each step.

_____ Understand: I circled key words.

_____ Plan: To solve the problem, I will _____.

_____ Solve: The answer is _____.

_____ Check: I checked my answer by _____.

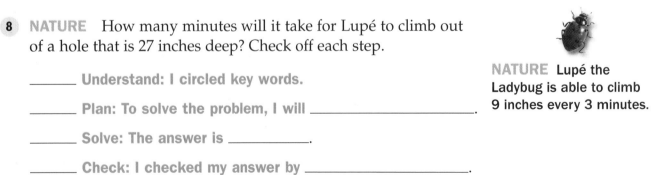

NATURE Lupé the Ladybug is able to climb 9 inches every 3 minutes.

9 FOOD   Kenyon and 5 friends went to the ice cream parlor. They each ordered 2 scoops of ice cream. How many scoops of ice cream were served to Kenyon and his 5 friends?

_____

10 **Reflect**   Explain two ways to find the number of ears on 5 students.

_____

_____

▶ Skills, Concepts, and Problem Solving

Write a possible situation for each rule.

11 Add 6 for each additional desk.

_____

12 Multiply by 9 for each additional tree.

_____

13 TRACK   Jeremy walked at a rate of 6 laps per hour on the track. How many laps did Jeremy walk in 4 hours?

The rule is add _____ laps for every _____ hour(s).

Jeremy walked _____ laps in 4 hours.

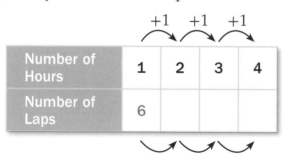

14 **TRAVEL**   Careta traveled at a rate of 70 miles for every 2 hours. How many miles did Careta travel in 8 hours? _____

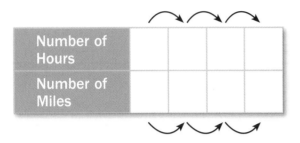

Number of Hours				
Number of Miles				

Solve.

15 **FOOD**   Sherita is selling lemonade for $1 per glass. In the first hour of business, she sold 10 glasses. How much money did Sherita make?

_____

A triangle has 3 sides.

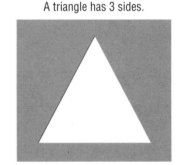

16 **GEOMETRY**   Aiden had to cut 7 triangles out of a sheet of construction paper. How many sides did Aiden cut out of the construction paper?

_____

17 **FOOD**   The chicken-finger appetizer at the Chick Pantry comes with 4 pieces of chicken. The cooks had 6 orders at the same time. How many pieces of chicken did they have to prepare?

_____

18 **SLEEP**   It is recommended that a person should get at least 8 hours of sleep each night. What is the least number of hours a person should sleep in 1 week?

_____

Vocabulary Check   **Write the vocabulary word that completes each sentence.**

19 A(n) _____ tells how numbers are related to each other.

20 A(n) _____ is a sequence of numbers, figures, or symbols that follows a rule or design.

21 **Writing in Math** Explain how to find the number of legs that 6 octopuses have by using the photo at the right.

_____

_____

An octopus has 8 legs.

 Spiral Review

Write each number in expanded form. (Lesson 1-1, p. 4)

22 908 _____

23 204 _____

24 1,367 _____

25 4,972 _____

Write each number in standard form. (Lesson 1-2, p. 9)

26 two hundred seven

_____

27 one million, forty-five

_____

28 two million, one hundred eighty

_____

29 nine hundred thousand, six hundred

_____

Solve. (Lesson 1-3, p. 16)

30 PICTURES Emil made a pyramid of pictures for a school project. He put 3 pictures in the top row, 6 pictures in the second row, and 9 pictures in the third row. If the pattern continues, how many pictures will be in the fifth row?

_____

STOP

Progress Check 2 (Lessons 1-3 and 1-4)

In each sequence, find a rule. Then write the next three terms.

1 315, 296, 277, 258, _____, _____, _____

Rule: _____

Next terms: _____, _____, _____

2 9, 18, 36, 72, _____, _____, _____

Rule: _____

Next terms: _____, _____, _____

Write the next three conversions in each pattern.

3

Number of Milliliters	1,000	2,000	3,000	4,000
Number of Liters	1			

4

Number of Hours	1	2	3	4
Number of Minutes	60			

Solve.

5 **HEALTH**   Tyree sleeps 9 hours each night. How many hours does he sleep in 1 week?

Tyree sleeps _____ hours in one week.

6 **MEASUREMENT**   One bucket can hold 8 gallons of water. How many gallons of water can 7 buckets hold?

Seven buckets can hold _____ gallons of water.

7 **FITNESS**   Steve exercises 0.75 hour each day. How many hours will he exercise in 8 days?

Steve will exercise _____ hours in 8 days.

8 **SHARKS**   Refer to the photo at the right. About how many inches does a great white shark grow in 7 years?

A great white shark grows about _____ inches in 7 years.

SHARKS Great white sharks grow about 10 inches per year.

Vocabulary and Concept Check

expanded form, *p. 4*

pattern, *p. 16*

period, *p. 9*

rule, *p. 16*

standard form, *p. 4*

word form, *p. 9*

Write the vocabulary word that completes each sentence.

1 A number is written in _____ using only digits.

2 A(n) _____ tells how the numbers, figures, or symbols in a pattern are related to each other.

3 A group of three digits in the place-value chart is known as a _____ .

4 _____ shows a number as a sum of the value of each digit.

Label each diagram below. Write the correct place value in each blank.

5 _____

23,496

6 _____

8,419,650

Lesson Review

1-1 Whole Numbers Less Than 10,000 (pp. 4-8)

Identify the value of each underlined digit.

7 4,853 _____

8 7,090 _____

9 9,248 _____

10 6,092 _____

11 3,381 _____

12 5,305 _____

> ### Example 1
>
> **Identify the value of the underlined digit in 3,506.**
>
> Write each digit in the place-value chart.
>
> The underlined digit is in the hundreds place.
>
> | 1000 | 100 | 10 | 1 |
> |------|-----|-----|-----|
> | thousands | hundreds | tens | ones |
> | 3 | 5 | O | 6 |
>
> Multiply the underlined digit by the value of its place. 5 × 100 = 500
>
> The underlined digit has a value of 500.

1-2 Read and Write Whole Numbers in the Millions (pp. 9-14)

Write each number in word form.

13 5,854,120

_____

_____

_____

14 258,764

_____

_____

_____

15 3,400,250

_____

16 14,273

_____

Example 2

Write 2,375,608 in word form.

Rewrite the number using digits and words (using period names).

Write the millions period. two million

Write the thousands period. three hundred seventy-five thousand

Write the ones period. six hundred eight

Write the periods together.
two million, three hundred seventy-five thousand, six hundred eight

1-3 Number Relationships (pp. 16-22)

What is the rule for each pattern?

17 185, 179, 173, 167 _____

18 4, 8, 16, 32 _____

19 2, 6, 18, 54 _____

Write the next three terms in the pattern.

20 18, 30, 42, 54, _____, _____, _____

21 3, 10, 17, _____, _____, _____

22 856, 736, 616, 496, _____, _____, _____

Example 3

Write the next three terms in the pattern.
2, 5, 11, 23, _____, _____, _____

Find the rule.

$2 \times 2 = 4 + 1 = 5$

$5 \times 2 = 10 + 1 = 11$

$11 \times 2 = 22 + 1 = 23$

The rule is multiply by 2, and then add 1.

Continue the pattern.

$23 \times 2 = 46 + 1 = 47$

$47 \times 2 = 94 + 1 = 95$

$95 \times 2 = 190 + 1 = 191$

The next three terms are 47, 95, 191.

Solve.

23 **SHAPES** A pentagon has 5 sides. How many sides do 7 pentagons have?

_____

24 **HOMEWORK** Regina needs to read 3 chapters this week. Each chapter has 23 pages. How many pages does Regina need to read?

_____

25 **TRAVEL** Cesar's dad drives 8 miles to work and 8 miles home 5 days a week. How many miles does his dad drive to and from work each week?

_____

Example 4

How many toes are on 6 feet?

Each foot has 5 toes.

One rule is "add 5 for each foot."
$$5 + 5 + 5 + 5 + 5 + 5 = 30$$

Another rule is "multiply the number of feet by 5."
$$6 \times 5 = 30$$

There are 30 toes on 6 feet.

1-4 Linear Patterns (pp. 23-29)

Solve.

26 A spider has 8 legs. How many legs do 4 spiders have?

_____

27 Joselyn can run 1 mile in 7 minutes. If she keeps a constant pace, how many miles can she run in 35 minutes?

_____

28 Ross and his 5 friends each bought 3 comic books. How many comic books did they buy in all?

_____

Example 5

Janice works at her father's business for $8 per hour. How much money will Janice earn if she works for 5 hours? Make a table.

For every 1 hour worked, Janice earns $8. So the amount of money earned increases by $8 for each additional hour. The rule is add 8 for each hour worked.

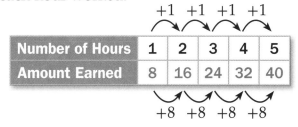

Number of Hours	1	2	3	4	5
Amount Earned	8	16	24	32	40

Janice will earn $40 for working 5 hours.

Write each number in standard form.

1   seven hundred forty-six thousand, two hundred eight _____

2   one hundred fifty-two thousand, six hundred eleven _____

3   2,000 + 800 + 50 + 9 _____

4   8,000 + 500 + 20 + 1 _____

Write each number in expanded form.

5   9,764 _____

6   2,505 _____

7   3,491 _____

8   6,700 _____

Identify the value of each underlined digit.

9   7,<u>8</u>90 _____

10   <u>3</u>,080 _____

11   <u>5</u>,492 _____

12   3,0<u>8</u>1 _____

13   6,9<u>1</u>9 _____

14   2,<u>3</u>04 _____

Write each number in word form.

15   2,378,490

_____

16   254,300

_____

17   3,511,780

_____

18   8,211,099

_____

Write the next three terms.

19 11, 18, 25, 32

20 3, 9, 27, 81

21 100, 85, 70, 55

_____, _____, _____   _____, _____, _____   _____, _____, _____

Write the next three conversions in each pattern.

22

Number of Yards	1	2	3	4
Number of Inches	36			

23

Number of Kilometers	1	2	3	4
Number of Meters	1,000			

State the rule. Then find the solution for each of the following.

24 How many tires do four "18-wheeler" tractor-trailer trucks have?

_____

25 How many legs do 7 cats have?

_____

Solve.

26 **TRAVEL**   During each day of her 1 week vacation in the mountains, Monica biked a distance of 15 miles. Over the entire vacation, how many miles did Monica bike?

_____

Correct the mistakes.

27 **BANKING**   When writing a check, you must write both the standard form and word form of the dollar amount of the check. Mr. Krueger made a purchase at Kid Creations for $375. He wrote the check shown below. Explain what Mr. Krueger did wrong.

_____

_____

Kelsey Krueger
123 High Street,
Santa Cruz, CA 95064

001

DATE _9/14/2007_

PAYEE ___Kid Creations___ $ _375.00_
___Three seventy-five___ / 00 DOLLARS

BANK ●   ___Kelsey Krueger___

"'001 "'12345"123'" 123456"123"'

STOP

Test Practice

Choose the best answer and fill in the corresponding circle on the sheet at right.

1 Which number shows nine thousand three written in standard form?

   A 9,003   **C** 9,300

   B 9,013   **D** 9,330

2 $7,000 + 60 + 2 =$

   A 762   **C** 7,602

   B 7,062   **D** 7,662

3 Which digit is in the thousands place in 7,005,012?

   A 0   **C** 5

   B 1   **D** 7

4 Which of these is the number 6,007,017?

   A six million, seven hundred, seventeen

   B six thousand, seven hundred, seventeen

   C six billion, seven million, seventeen

   D six million, seven thousand, seventeen

5 What is the next number in the sequence?

       9, 18, 27, 36, 45, _____

   A 50   **C** 58

   B 54   **D** 60

6 Samson can run 2 miles in 10 minutes. If he keeps this pace, how long will it take him to run 4 miles?

   A 5 minutes   **C** 15 minutes

   B 10 minutes   **D** 20 minutes

7 What is the rule for this pattern?

       22, 19, 16, 13, 10, 7, 4, 1

   A add 4   **C** subtract 4

   B subtract 3   **D** add 5

8 Ducks have 2 legs. Rico is feeding ducks at the pond. He counts 28 duck legs. How many ducks are there?

   A 7 ducks   **C** 14 ducks

   B 28 ducks   **D** 56 ducks

9 The numbers in the pattern increase by the same amount each time. What are the next three numbers in the pattern?

     23, 40, 57, 74, _____, _____, _____

   A 99, 124, 149

   B 86, 98, 110

   C 91, 108, 125

   D 93, 112, 131

GO ON

10 Paige's family is gathered in the living room for family game night. She has a large family and a few dogs. If there is a total of 24 legs in this room, how many humans and how many dogs are there?

A 6 humans, 3 dogs

B 4 humans, 2 dogs

C 3 humans, 6 dogs

D 5 humans, 4 dogs

11 Stacey bought 5 bottles of hairspray. Each bottle of hairspray cost $2. How much did Stacey spend?

A $8

B $12

C $10

D $14

12 Valerie bought a watch that was 4 times as much as the hat Bea bought. Bea spent $22. How much did Valerie spend?

A $80

B $88

C $84

D $100

ANSWER SHEET

Directions: Fill in the circle of each correct answer.

1   Ⓐ Ⓑ Ⓒ Ⓓ
2   Ⓐ Ⓑ Ⓒ Ⓓ
3   Ⓐ Ⓑ Ⓒ Ⓓ
4   Ⓐ Ⓑ Ⓒ Ⓓ
5   Ⓐ Ⓑ Ⓒ Ⓓ
6   Ⓐ Ⓑ Ⓒ Ⓓ
7   Ⓐ Ⓑ Ⓒ Ⓓ
8   Ⓐ Ⓑ Ⓒ Ⓓ
9   Ⓐ Ⓑ Ⓒ Ⓓ
10 Ⓐ Ⓑ Ⓒ Ⓓ
11 Ⓐ Ⓑ Ⓒ Ⓓ
12 Ⓐ Ⓑ Ⓒ Ⓓ

Success Strategy

Double check your answers after you finish. Read each problem and all of the answer choices. Put your finger on each bubble you filled in to make sure it matches the answer for each problem.

Chapter 2
Multiplication

How do you use multiplication?

Melanie and her mom want to buy five pumpkins. Suppose each pumpkin costs $3.00. You can use multiplication to quickly find the cost of multiple items.

STEP **2** Preview   Get ready for Chapter 2. Review these skills and compare them with what you'll learn in this chapter.

What You Know	What You Will Learn
You know how to add. **Examples:** $5 + 5 + 5 + 5 = 20$ $2 + 2 + 2 = 6$ **TRY IT!** **1** $3 + 3 =$ _____ **2** $10 + 10 + 10 + 10 =$ _____ **3** $6 + 6 + 6 =$ _____ **4** $4 + 4 + 4 + 4 =$ _____	*Lessons 2-1 and 2-6* **Multiplication** is repeated addition. $5 + 5 + 5 + 5 = 5 \times 4$ You can use **arrays** to model multiplication. $2 \times 3 =$
You know how to skip count. **Example:** Skip count by 5s. $0, 5, 10, 15, 20, 25, 30, 35, 40, 45, 50,\ldots$ **TRY IT!** **5** Skip count by 4s. _____ **6** Skip count by 6s. _____	*Lesson 2-3* **Multiples** of 8 are the numbers you say when you skip count by 8s. $0, 8, 16, 24, 32, 40, 48, 56, 64, 72, \ldots$ The multiples of 8 are the multiplication facts below. $0 \times 8 = 0 \qquad 5 \times 8 = 40$ $1 \times 8 = 8 \qquad 6 \times 8 = 48$ $2 \times 8 = 16 \qquad 7 \times 8 = 56$ $3 \times 8 = 24 \qquad 8 \times 8 = 64$ $4 \times 8 = 32 \qquad 9 \times 8 = 72$ $\qquad\qquad\qquad 10 \times 8 = 80$
You know that addition and subtraction are inverse operations. $15 - 7 = 8 \qquad 8 + 7 = 15$	*Lesson 2-6* **Multiplication** and **division** are also **inverse operations**. They undo each other. $8 \times 5 = 40 \qquad 40 \div 5 = 8$

Multiply by 0, 1, 5, and 10

KEY Concept

The **Zero Property of Multiplication** states that any number multiplied by zero is zero.

$12 \times 0 = 0$ because 12 groups of zero is zero.

The **Identity Property of Multiplication** states that any number multiplied by 1 is equal to that number.

$12 \times 1 = 12$ because 12 groups of 1 is 12.

Skip count by 5s to say the multiples of 5.

1   2   3   4   5   6   7   8   9   10
5, 10, 15, 20, 25, 30, 35, 40, 45, 50

15 is the third multiple of 5, so $3 \times 5 = 15$.

Skip count by 10s to say the multiples of 10.

$1 \times 10 = 10$	$2 \times 10 = 20$	$3 \times 10 = 30$
$4 \times 10 = 40$	$5 \times 10 = 50$	$6 \times 10 = 60$
$7 \times 10 = 70$	$8 \times 10 = 80$	$9 \times 10 = 90$

$10 \times 10 = 100$

To multiply by 10, write a 0 to the place on the right of the number.

VOCABULARY

factor
a number that divides into a whole number evenly; also a number that is multiplied by another number

Identity Property of Multiplication
when a number is multiplied by 1, the product is the same as the given number
Example: $8 \times 1 = 8$

multiplication
an operation on two numbers to find their product; it can be thought of as repeated addition

product
the answer or result of a multiplication problem; it also refers to expressing a number as the product of its factors

Zero Property of Multiplication
property that states any number multiplied by zero is zero

The Identity and Zero Properties hold true for numbers of any size. Use skip counting to learn multiplication facts with 5 or 10 as factors.

Example 1

Find the product of 8 and 0.

1. What is the first factor?   8

   What is the second factor?   0

2. Use the Zero Property of Multiplication. Any number multiplied by zero is zero.

3. Write the product.
   $8 \times 0 = 0$

YOUR TURN!

Find the product of 2 and 0.

1. What is the first factor?   _____

   What is the second factor? _____

2. Use the Zero Property of Multiplication.

3. Write the product.

   _____

Example 2

Find the product of 46 and 1.

1. What is the first factor? **46**

   What is the second factor? **1**

2. Use the Identity Property of Multiplication. Any number multiplied by 1 is equal to the given number.

3. Write the product.
   46 × 1 = 46

Find the product of 69 and 1.

1. What is the first factor? _____

   What is the second factor? _____

2. Use the Identity Property of Multiplication. Any number multiplied by _____ is equal to the given number.

3. Write the product.

   _____

Example 3

Find the product of 10 and 73.

1. Use the rule for multiplying by 10.

   To multiply by 10, write a 0 to the place on the right of the number.

2. 10 × 73 = **730**

3. Estimate to check.

   73 is close to 70.

   10 × 70 = **700**

   700 is close to 730, so the answer makes sense.

Find the product of 10 and 16.

1. Use the rule for multiplying by 10.

   To multiply by 10, write a _____ to the place on the right of the number.

2. 10 × 16 = _____

3. Estimate to check.

   16 is close to _____.

   10 × _____ = _____

   _____ is close to _____, so the answer makes sense.

GO ON

Example 4

Find the product of 5 and 15.

1. Rewrite the problem in a vertical format.

2. Multiply the number in the ones column by 5.   **5 × 5 = 25**
   Write the tens digit above the tens column.
   Write the ones digit under the ones column as part of the product.

$$\begin{array}{r} {\scriptstyle 2} \\ 15 \\ \times\ 5 \\ \hline 5 \end{array}$$

3. Multiply 5 times the digit in the tens column.   **5 × 1 = 5**
   Add the 2 regrouped tens for a total of **7** tens.
   The product is 75.

$$\begin{array}{r} {\scriptstyle 2} \\ 15 \\ \times\ 5 \\ \hline 75 \end{array}$$

4. Skip count to check. 15 × 5 is the 15th multiple of 5.

     1   2   3   4   5   6   7   8   9 10 11 12 13 14 15
   5, 10, 15, 20, 25, 30, 35, 40, 45, 50, 55, 60, 65, 70, 75

The 15th multiple of 5 is 75, so the answer makes sense.

YOUR TURN!

Find the product of 5 and 32.

1. Rewrite the problem in a vertical format.

2. Multiply the number in the ones column by 5. 5 × 2 = _____
   Write the tens digit above the tens column.
   Write the ones digit under the ones column as part of the product.

$$\begin{array}{r} 32 \\ \times\ 5 \\ \hline \end{array}$$

3. Multiply 5 times the digit in the tens column. 5 × 3 = _____
   Add in the 1 regrouped ten for a total of _____ tens.
   The product is _____.

$$\begin{array}{r} {\scriptstyle 1} \\ 32 \\ \times\ 5 \\ \hline 0 \end{array}$$

4. Skip count to check. 32 × 5 is the _____ multiple of 5.

   5, 10, 15, 20, 25,…   145, 150, _____, _____
   The 32nd multiple of 5 is _____, so the answer makes sense.

Who is Correct?

Find the product of 10 and 21.

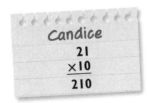

Candice	Malcolm	Liana
21	21	21
×10	×10	×10
210	201	2,100

Circle correct answer(s). Cross out incorrect answer(s).

Guided Practice

Use the Zero Property or the Identity Property of Multiplication to find each product.

1 Find the product of 62 × 0.

What is the first factor? _____

What is the second factor? _____

Which property should you use?

_____

Write the product. _____

2 Find the product of 832 × 1.

What is the first factor? _____

What is the second factor? _____

Which property should you use?

_____

Write the product. _____

Step by Step Practice

3 Find the product of 5 and 407.

Step 1 Rewrite the problem in a vertical format.

Step 2 Multiply the number in the ones column by 5.
5 × 7 = _____
Write the tens digit above the tens column.
Write the ones digit under the ones column as part of the product.

$$\begin{array}{r} 407 \\ \times\ 5 \\ \hline \end{array}$$

Step 3 Multiply 5 times the digit in the tens column.
5 × 0 = _____
Add the 3 regrouped tens for a total of 3 tens.

$$\begin{array}{r} 407 \\ \times\ 5 \\ \hline \end{array}$$

Step 4 Multiply 5 times the digit in the hundreds column.

5 × 400 = _____

$$\begin{array}{r} 407 \\ \times\ 5 \\ \hline \end{array}$$

Step 5 5 × 407 = _____

GO ON

Find each product. Show your work.

4 $193 \times 0 =$ _____

5 $774 \times 1 =$ _____

6 $10 \times 86 =$ _____

7 $5 \times 34 =$ _____

8 $1 \times 15 =$ _____

9 $0 \times 42 =$ _____

10 $5 \times 20 =$ _____

11 $10 \times 30 =$ _____

Step by Step Problem-Solving Practice

Solve.

12 **HOUSING** A housing development has 5 houses on each acre of land. The development sits on 26 acres of land. How many houses are in the development?

Understand Read the problem. Write what you know.

There are _____ houses on each acre of land.

There are _____ acres of land.

Plan Pick a strategy. One strategy is to solve a simpler problem.

Solve You need to find the product of 5 and 26. Think: $26 = 20 + 6$. Multiply 5×20 and 5×6.

Multiply 5 times the ones place value.

Multiply 5 times the tens place value.

Add the products.

A total of _____ houses are built.

Check You can multiply 5×26 vertically to check.

13 **SCHOOL** Tara is passing out one pencil to each student before a test. There are 10 rows of students. Each row has 8 students. How many pencils will be passed out?

Check off each step.

_____ **Understand: I circled key words.**

_____ **Plan: To solve the problem, I will** _____.

_____ **Solve: The answer is** _____.

_____ **Check: I checked my answer by** _____.

14 **PETS** Alberto's pet hamster eats 5 food pellets each day. How many food pellets will Alberto's hamster eat in 30 days?

_____

15 **Reflect** How is the product of a number multiplied by 10 similar to the number itself?

_____

_____

▶ Skills, Concepts, and Problem Solving

Find each product. Show your work.

16 $8 \times 1 =$ _____

17 $6 \times 0 =$ _____

18 $5 \times 6 =$ _____

19 $10 \times 42 =$ _____

20 $5 \times 36 =$ _____

21 $1 \times 50 =$ _____

22 $52 \times 5 =$ _____

23 $0 \times 100 =$ _____

24 $1 \times 200 =$ _____

25 $10 \times 81 =$ _____

26 $5 \times 97 =$ _____

27 $122 \times 1 =$ _____

GO ON

Find each product. Show your work.

28 26 × 0 = _____

29 98 × 1 = _____

30 14 × 5 = _____

31 73 × 10 = _____

32 1 × 573 = _____

33 10 × 206 = _____

34 0 × 841 = _____

35 5 × 190 = _____

36 45 × 1 = _____

Solve.

37 PIANO   Julius practices the piano for 45 minutes each day.
How many minutes will Julius practice in 10 days?

_____

38 SCHOOL SUPPLIES   Ying is buying school supplies. She buys
5 packs of paper. There are 250 pages of paper in each pack.
How many pages of paper did Ying buy?

_____

Vocabulary Check   **Write the vocabulary word that completes each sentence.**

39 _____ can be thought of as repeated addition.

40 The answer or result of a multiplication problem is the _____.

41 Writing in Math   When might you need to multiply a number
by zero? Give an example and explain.

_____

_____

_____

Multiply by 2, 3, 4, and 6

KEY Concept

Multiples are the numbers you say when you skip count.

- Multiples of 2: 2, 4, 6, 8, 10, 12, 14, 16, 18, 20
- Multiples of 3: 3, 6, 9, 12, 15, 18, 21, 24, 27, 30
- Multiples of 4: 4, 8, 12, 16, 20, 24, 28, 32, 36, 40
- Multiples of 6: 6, 12, 18, 24, 30, 36, 42, 48, 54, 60

The **Commutative Property of Multiplication** states that the order of factors does not change the product.

$$4 \times 6 = 24 \qquad 6 \times 4 = 24$$

You can find 25×6 in different ways.

$$
\begin{array}{r}
20 + 5 \\
\times \quad 6 \\
\hline
120 + 30 \\
\end{array}
\qquad
\begin{array}{r}
\overset{3}{25} \\
\times\, 6 \\
\hline
150 \\
\end{array}
$$

$$150$$

You should practice memorizing the multiplication facts of 2, 3, 4, and 6.

VOCABULARY

array
objects or symbols displayed in rows of the same length and columns of the same length; the length of a row might be different from the length of a column

Commutative Property of Multiplication
the order in which two numbers are multiplied does not change the product

factor
a number that divides into a whole number evenly; also a number that is multiplied by another number

multiplication
an operation on two numbers to find their product; it can be thought of as repeated addition

product
the answer or result of a multiplication problem; it also refers to expressing a number as the product of its factors

Example 1

Draw an array to model the expression 2×9. Find the product.

1. The first factor is 2, so there will be 2 rows. The second factor is 9, so there will be 9 columns, or 9 in each row.

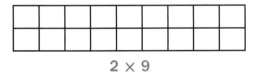

2×9

2. Label the array 2×9. Count the rectangles. 18

3. Write the multiplication fact. $2 \times 9 = 18$

GO ON

YOUR TURN!

Draw an array to model the expression 3 × 8. Find the product.

1. The first factor is _____, so there will be _____ rows.
   The second factor is _____, so there will be _____ columns.

2. Label the array 3 × 8. There are _____ rectangles.

3. Write the multiplication fact. _____ × _____ = _____

Example 2

Find the product of 3 and 59.

1. Rewrite the problem in a vertical format.

2. Multiply the number in the ones column by 3.   3 × 9 = 27
   Write the ones digit under the ones column as part of the product.
   Write the tens digit above the tens column.

$$\begin{array}{r} {}^{2} \\ 59 \\ \times\ 3 \\ \hline 7 \end{array}$$

3. Multiply 3 times the digit in the tens column.   3 × 5 = 15
   Add the 2 regrouped tens for a total of **17** tens.

4. 3 × 59 = 177

$$\begin{array}{r} {}^{2} \\ 59 \\ \times\ 3 \\ \hline 177 \end{array}$$

YOUR TURN!

Find the product of 4 and 37.

1. Rewrite the problem in a vertical format.

2. Multiply the number in the ones column by 4.
   Write the ones digit under the ones column as part of the product.
   Write the tens digit above the tens column.

$$\begin{array}{r} 37 \\ \times\ 4 \\ \hline \end{array}$$

3. Multiply 4 times the digit in the tens column.
   4 × 3 = _____
   Add the 2 regrouped tens for a total of _____ tens.

4. 4 × 37 = _____

$$\begin{array}{r} {}^{2} \\ 37 \\ \times\ 4 \\ \hline 8 \end{array}$$

Who is Correct?

Find the product of 83 and 6.

Sandy
83
×6
4,818

Alvin
83
×6
488

Kiyo
83
×6
498

Circle correct answer(s). Cross out incorrect answer(s).

 Guided Practice

Draw an array to model each expression. Find each product.

1 $3 \times 6 =$ _____

2 $2 \times 4 =$ _____

Step by Step Practice

3 Find the product of 77 and 6.

Step 1 Rewrite the problem in a vertical format.

Step 2 Multiply the number in the ones column by 6. $6 \times 7 =$ _____
Write the tens digit above the tens column.
Write the ones digit under the ones column as part of the product.

$$\begin{array}{r} 77 \\ \times\, 6 \\ \hline \end{array}$$

Step 3 Multiply 6 times the digit in the tens column.
$6 \times 7 =$ _____
Add the 4 regrouped tens for a total of 46 tens.

$$\begin{array}{r} 77 \\ \times\, 6 \\ \hline \end{array}$$

Step 4 The product is 462.

 GO ON

Find each product. Show your work.

4 $12 \times 2 =$ _____

5 $9 \times 3 =$ _____

6 $7 \times 4 =$ _____

7 $6 \times 6 =$ _____

8 $2 \times 25 =$ _____

9 $3 \times 16 =$ _____

10 $4 \times 51 =$ _____

11 $6 \times 33 =$ _____

12 $3 \times 21 =$ _____

Step by Step *Problem-Solving Practice*

Solve.

13 **CARS** A car factory is shipped 19 new car bodies for assembly. With 4 tires needed per car, how many tires are needed to assemble the entire shipment?

Problem-Solving Strategies
☐ Draw a model.
☑ Use a logical reasoning.
☐ Make a table.
☐ Solve a simpler problem.
☐ Work backward.

Understand   Read the problem. Write what you know.

There are _____ tires for each car.

There were _____ car bodies shipped.

Plan   Pick a strategy. One strategy is to use logical reasoning.

Solve   Think: 19 is close to 20.

$4 \times 20 =$ _____, so the product

should be close to _____.

$$\begin{array}{r} 19 \\ \times\ 4 \\ \hline \end{array}$$

_____ tires are needed to assemble the entire shipment.

Check   Compare the answer to the estimate.

14 **BIRDS** There are 12 bird nests in a tree. There are 6 eggs in each nest. What is the total number of eggs in the tree?

Check off each step.

_____ Understand: I circled key words.

_____ Plan: To solve the problem, I will _____.

_____ Solve: The answer is _____.

_____ Check: I checked my answer by _____.

15 **SCHOOL** Mrs. Romero gives stickers to her students when they do well on exams. She estimates that she will need 4 stickers for each student. There are 33 students in Mrs. Romero's class. How many stickers will Mrs. Romero need?

_____

16 **Reflect** How is multiplication like repeated addition? Give an example and explain.

_____

_____

_____

Skills, Concepts, and Problem Solving

Draw an array to model each expression. Find each product.

17 4 × 5 = _____

18 2 × 5 = _____

19 6 × 3 = _____

20 3 × 4 = _____

GO ON

Draw an array to model each expression. Find each product.

21 2 × 4 = _____

22 4 × 4 = _____

23 3 × 3 = _____

24 2 × 6 = _____

Find each product. Show your work.

25 6 × 4 = _____

26 3 × 2 = _____

27 8 × 2 = _____

28 4 × 23 = _____

29 3 × 46 = _____

30 6 × 63 = _____

31 2 × 98 = _____

32 73 × 3 = _____

33 19 × 4 = _____

34 6 × 200 = _____

35 3 × 40 = _____

36 55 × 2 = _____

37 4 × 493 = _____

38 222 × 6 = _____

39 2 × 99 = _____

40 BICYCLES Mr. Wilson's bicycle shop has 234 bicycles in stock. Each bicycle has 2 wheels. How many wheels are in Mr. Wilson's bicycle shop?

_____

41 TREES Each day, 315 trees are planted in Westin Woods. How many trees will be planted in 6 days?

_____

Vocabulary Check **Write the vocabulary word that completes each sentence.**

42 _____ is an operation on two numbers to find their product.

43 A(n) _____ is objects or symbols displayed in rows of the same length and columns of the same length.

44 Writing in Math Explain how to use the Commutative Property to write $6 \times 2 \times 3$ three different ways. Solve the problem.

_____

_____

_____

▶ **Spiral Review** (Lesson 2-1, p. 40)

Solve.

45 $5 \times 264 =$ _____

46 $0 \times 26 =$ _____

47 57×10 _____

48 NEWSPAPERS Kerri is delivering newspapers on her paper route. There are 72 houses on Kerri's route. Each house gets 1 newspaper. How many newspapers will Kerri deliver?

_____

Find each product. Show your work.

1 5 × 7 = _____

2 10 × 4 = _____

3 2 × 9 = _____

4 3 × 4 = _____

5 4 × 6 = _____

6 6 × 8 = _____

7 29 × 0 = _____

8 98 × 1 = _____

9 5 × 41 = _____

10 10 × 17 = _____

11 19 × 2 = _____

12 27 × 3 = _____

13 36 × 4 = _____

14 71 × 6 = _____

15 92 × 10 = _____

Use the Identity Property of Multiplication to solve.

16 32 × _____ = _____

17 _____ × 192 = _____

Use the Zero Property of Multiplication to solve.

18 418 × _____ = _____

19 _____ × 22 = _____

Solve.

20 **LUNCH**   Tamara eats 5 grapes every day for lunch. How many grapes will Tamara eat in 97 days?

_____

21 **SCHOOL**   There are 10 students in Mr. Castillo's class. Each student has 10 fingers. How many fingers do these students have altogether?

_____

Multiply by 7, 8, and 9

KEY Concept

Multiples are the numbers you say when you skip count.

- Multiples of 7: 7, 14, 21, 28, 35, 42, 49, 56, 63, 70
- Multiples of 8: 8, 16, 24, 32, 40, 48, 56, 64, 72, 80
- Multiples of 9: 9, 18, 27, 36, 45, 54, 63, 72, 81, 90

You can find 35×8 in several ways.

$$\begin{array}{cc} 30 & 5 \\ \times 8 & \times 8 \\ \hline 240 & 40 \end{array} \qquad \begin{array}{c} 30 + 5 \\ \times \quad 8 \\ \hline 240 + 40 \end{array} \qquad \begin{array}{c} \overset{4}{35} \\ \times 8 \\ \hline 280 \end{array}$$

$$280 \qquad\qquad 280$$

Practice memorizing the multiplication facts of 7, 8, and 9. This will make multiplying two- and three-digit numbers easier.

VOCABULARY

factor
  a number that divides into a whole number evenly; also a number that is multiplied by another number

multiple
  a multiple of a number is the product of that number and any whole number.
  Example: 30 is a multiple of 10 because $3 \times 10 = 30$.

product
  the answer or result of a multiplication problem; it also refers to expressing a number as the product of its factors

Example 1

Use a pattern to find the product of 9×9.

1. Skip count by 9s to write the multiples of 9. Facts with 9 make a pattern.

2. Multiples of 9:
   9, 18, 27, 36, 45, 54, 63, 72, 81, 90

   $1 \times 9 = 9$ $6 \times 9 = 54$
   $2 \times 9 = 18$ $7 \times 9 = 63$
   $3 \times 9 = 27$ $8 \times 9 = 72$
   $4 \times 9 = 36$ $9 \times 9 = 81$
   $5 \times 9 = 45$ $10 \times 9 = 90$

3. $9 \times 9 = $ **81**

YOUR TURN!

Use a pattern to find the product of 6×7.

1. Skip count by 7s to write the multiples of 7.

2. Multiples of 7: _____, _____, _____,
   _____, _____, _____

   $1 \times 7 = $ _____ $6 \times 7 = $ _____
   $2 \times 7 = $ _____ $7 \times 7 = $ _____
   $3 \times 7 = $ _____ $8 \times 7 = $ _____
   $4 \times 7 = $ _____ $9 \times 7 = $ _____
   $5 \times 7 = $ _____ $10 \times 7 = $ _____

3. $6 \times 7 = $ _____

GO ON

Example 2

Find the product of 8 and 35. Use doubling.

Double 1 and you have 2.
Double 2 and you have 4.
Double 4 and you have 8.
You can find the product of any number and 8 by doubling the number three times.

1. Double 35.

$$\begin{array}{r} \overset{1}{35} \\ \times\ 2 \\ \hline 70 \end{array}$$

2. Double the product.

$$\begin{array}{r} 70 \\ \times\ 2 \\ \hline 140 \end{array}$$

3. Double the product again.

$$\begin{array}{r} 140 \\ \times\ 2 \\ \hline 280 \end{array}$$

$8 \times 35 = 280$

YOUR TURN!

Find the product of 8 and 49. Use doubling.

Double 1 and you have 2.

Double 2 and you have _____.

Double 4 and you have _____.

1. Double _____.

$$\begin{array}{r} 49 \\ \times\ 2 \end{array}$$

2. Double the product.

$$\begin{array}{r} 98 \\ \times\ 2 \end{array}$$

3. Double the product again.

$8 \times 49 =$ _____

$$\begin{array}{r} 196 \\ \times\ 2 \end{array}$$

Who is Correct?

Find the product of 24 and 9.

Lisa

$$\begin{array}{r} 24 \\ \times\ 9 \\ \hline 36 \\ +\ 18 \\ \hline 54 \end{array}$$

Ramon

$$\begin{array}{r} 24 \\ \times\ 9 \\ \hline 1,836 \end{array}$$

Kenisha

$$\begin{array}{r} \overset{3}{24} \\ \times\ 9 \\ \hline 216 \end{array}$$

Circle correct answer(s). Cross out incorrect answer(s).

 Guided Practice

Use a pattern to find each product.

1 $6 \times 8 =$ _____
   Multiples of 8: _____, _____, _____, _____, _____, _____, _____, _____, _____, _____

2 $7 \times 7 =$ _____
   Multiples of 7: _____, _____, _____, _____, _____, _____, _____, _____, _____, _____

Step by Step Practice

3 Find the product of 109 × 8. Use doubling.

Step 1 Double 109.

Step 2 Double the product.

Step 3 Double the product again.

Step 4 109 × 8 = _____

Find each product. Show your work.

4 7 × 5 = _____

5 8 × 7 = _____

6 9 × 6 = _____

7 14 × 7 = _____

8 22 × 8 = _____

9 30 × 9 = _____

10 7 × 501 = _____

11 8 × 234 = _____

12 9 × 111 = _____

GO ON

Solve.

13 **BASKETBALL** The Rockville City basketball league ends their season with an awards banquet. There are 14 teams in the league, each with 9 players. How many players will be invited to the awards banquet?

Understand	Read the problem. Write what you know.
	There are _____ basketball teams.
	There are _____ players on each team.
Plan	Pick a strategy. One strategy is to make a table. Make a table with two rows. Title one row "teams" and the other row "players."
Solve	Write 1 through 14 for teams, because there are 14 teams in the league. There are 9 players on each team, so add 9 each time.

Teams	1	2	3	4	5	6								
Players	9	18	27											

Check	Multiply 14 × 9 to check your answer.

14 **SCHOOL** The lunch room at Central School has 7 tables. Eight students can sit at each table. If all the seats are filled, how many students will be sitting in the lunch room?

Check off each step.

_____ **Understand: I circled key words.**

_____ **Plan: To solve this problem, I will** _____.

_____ **Solve: The answer is** _____.

_____ **Check: I checked my answer by** _____.

15 **WALKING** Nawat walks 4 miles per hour. If Nawat walks for 7 hours, how far will he have walked? Show your work.

_____

16 **Reflect** Explain the Commutative Property. Use the example $7 \times 8 = 56$ in your explanation.

_____

_____

_____

▶ Skills, Concepts, and Problem Solving

Use patterns to find each product.

17 $7 \times 10 =$ _____
Multiples of 7: _____, _____, _____, _____, _____, _____, _____, _____, _____, _____

18 $8 \times 8 =$ _____
Multiples of 8: _____, _____, _____, _____, _____, _____, _____, _____, _____, _____

19 $9 \times 9 =$ _____
Multiples of 9: _____, _____, _____, _____, _____, _____, _____, _____, _____, _____

Find each product. Show your work.

20 $7 \times 3 =$ _____

21 $8 \times 5 =$ _____

22 $9 \times 4 =$ _____

23 $2 \times 7 =$ _____

24 $6 \times 8 =$ _____

25 $3 \times 9 =$ _____

26 $9 \times 20 =$ _____

27 $7 \times 11 =$ _____

28 $8 \times 26 =$ _____

29 $49 \times 8 =$ _____

30 $29 \times 7 =$ _____

31 $99 \times 9 =$ _____

32 $880 \times 8 =$ _____

33 $211 \times 7 =$ _____

34 $6 \times 109 =$ _____

GO ON

35 **BABYSITTING** Camellia earns $8 an hour babysitting. If Camellia works for 6 hours, how much will she have earned?

_____

36 **BAKING** Al's bakery has 35 baskets full of rolls. Each basket has 9 rolls inside. How many rolls are there all together?

_____

Vocabulary Check **Write the vocabulary word or words that complete each sentence.**

37 A(n) _____ of a number is the product of that number and any whole number.

38 The answer or result of a multiplication problem is called the

_____.

39 **Writing in Math** Write the multiples of 9 from 1 to 10. Explain any pattern in the products.

_____

_____

_____

 Spiral Review (Lessons 2-1, p. 40 and 2-2, p. 47)

Find each product.

40 $5 \times 44 =$ _____

41 $0 \times 999 =$ _____

42 $92 \times 3 =$ _____

Solve. (Lesson 2-2, p. 47)

43 **BASEBALL CARDS** Luther bought 6 packs of baseball trading cards. There are 4 cards in each pack. How many cards did Luther buy?

_____

STOP

Multiply by 11 and 12

KEY Concept

When multiplying by 11 and 12, you can use **patterns** and **array** models.

Multiply by 11		
1	× 11 =	11
2	× 11 =	22
3	× 11 =	33
4	× 11 =	44
5	× 11 =	55
6	× 11 =	66
7	× 11 =	77
8	× 11 =	88
9	× 11 =	99

The table shows that when a single-digit number is multiplied by 11, the **product** is the digit repeated.

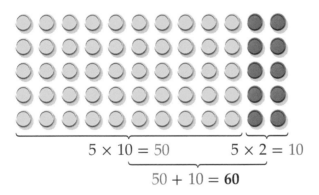

$5 \times 10 = 50$ $5 \times 2 = 10$

$50 + 10 = \mathbf{60}$

The array model shows that you can think of 12×5 as $(10 \times 5) + (2 \times 5)$. This is the **Distributive Property of Multiplication**.

VOCABULARY

array
> a display of objects or symbols in rows of the same length and columns of the same length

Distributive Property of Multiplication
> to multiply a sum by a number, multiply each addend by the number and add the products

pattern(s)
> a sequence of numbers, figures, or symbols that follows a rule or design

product
> the answer to a multiplication problem; it also refers to expressing a number as the product of its factors

Use array models, patterns, or the Distributive Property of Multiplication to multiply by 11 and 12.

Example 1

Use a pattern to find 8 × 11.

1. Use a table to complete the pattern.

1	× 11 =	11
2	× 11 =	22
3	× 11 =	33
4	× 11 =	44
5	× 11 =	55
6	× 11 =	66
7	× 11 =	77
8	× 11 =	88

2. 8 × 11 = **88**

YOUR TURN!

Use a pattern to find 8 × 12.

1. Use a table to complete the pattern.

1	× 12 =	12
2	× 12 =	24
3	× 12 =	36
4	× 12 =	48
5		
6		
7		
8		

2. 8 × 12 = _____

Example 2

Use an array model to find 12 × 7.

1. Rewrite 12 × 7 as (**10** × **7**) + (**2** × **7**).
   Then use the Distributive Property.

2. Draw an array using two sets of counters.

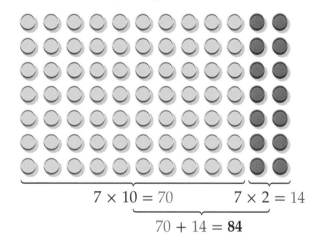

$$7 \times 10 = 70 \qquad 7 \times 2 = 14$$
$$70 + 14 = \mathbf{84}$$

3. The array shows 10 × 7 = 70 and
   2 × 7 = 14. So, 70 + 14 = 84.

4. 12 × 7 = **84**

YOUR TURN!

Use an array model to find 12 × 4.

1. Rewrite 12 × 4.

   (_____ × _____) + (_____ × _____)

2. Draw an array using two sets of counters.

3. The array shows 10 × _____ = _____
   and 2 × _____ = _____.
   So, _____ + _____ = _____.

4. 12 × 4 = _____

Who is Correct?

Find the product of 11 and 10.

Irena
(11 × 10) =
(10 × 10) + (1 × 10)
= 100 + 10
= 110

Amos
10 + 0
×10 + 1
─────
100 + 1
= 101

Gabe
11 × 10 =
(10 × 10) + (1 × 10)
= 100 + 100
= 200

Circle correct answer(s). Cross out incorrect answer(s).

Guided Practice

Rewrite the factors in distributive form.

1   12 × 3

2   12 × 6

Step by Step Practice

3   Use an array model to find 11 × 5.

Step 1   Rewrite 11 × 5.

(_____ × _____) + (1 × _____)

Step 2   Draw an array using two sets of counters.

Step 3   11 × 5 = (☐ × ☐) + (☐ × ☐)

= ☐ + ☐

= ☐

GO ON

Use an array model to find the product.

4 $12 \times 9 =$ _____

5 $11 \times 12 =$ _____

Find each product. Show your work.

6 $8 \times 11 =$ _____

7 $6 \times 12 =$ _____

8 $12 \times 10 =$ _____

9 $11 \times 7 =$ _____

Step by Step Problem-Solving Practice

Solve.

10 **INTERIOR DESIGN** Derrick's room is 11 feet by 10 feet. How many square feet of carpeting does Derrick need to carpet his room?

Understand	Read the problem. Write what you know.
	The room is _____ by _____ feet.
Plan	Pick a strategy. One strategy is solve a simpler problem.
Solve	Write 11×10 using the Distributive Property.
	(_____ × _____) + (_____ × _____)
	Add the products. _____
	Derrick needs _____ of carpet.
Check	You can skip count by 10s to the 11th number.

Problem-Solving Strategies
☐ Draw a picture.
☐ Use logical reasoning.
☑ Solve a simpler problem.
☐ Work backward.
☐ Make a table.

11 HOBBIES   There were 12 beads in a handmade bracelet. If Carla made 3 bracelets, how many beads did she use? Check off each step.

_____ Understand: I circled key words.

_____ Plan: To solve the problem, I will _____.

_____ Solve: The answer is _____.

_____ Check: I checked my answer by _____.

12 Reflect   Why does the pattern for multiplying by 11 only work for single digits?

_____

_____

 ## Skills, Concepts, and Problem Solving

Use patterns or array models to find the product.

13 $11 \times 9 =$ _____

14 $12 \times 5 =$ _____

15 $12 \times 7 =$ _____

16 $11 \times 7 =$ _____

GO ON

Find each product. Show your work.

17 12 × 11 = _____

18 1 × 11 = _____

19 12 × 2 = _____

20 11 × 10 = _____

21 11 × 6 = _____

22 12 × 4 = _____

23 11 × 11 = _____

24 12 × 3 = _____

Solve.

25 ASTRONOMY   Eli spent 12 minutes counting the stars in the sky. If he counted 12 stars each minute for 12 minutes, how many stars did he count?

_____

26 PHOTOS   Moesha developed 9 rolls of 12-exposure film. How many pictures did she develop?

_____

Vocabulary Check   **Write the vocabulary word that completes each sentence.**

27 A(n) _____ follows a rule or design.

28 Writing in Math   Explain how the word *dozen* can help you multiply by 12.

_____

_____

 Spiral Review

Find each product. (Lesson 2-3, p. 55)

29 7 × 4 = _____

30 8 × 9 = _____

Solve. (Lesson 2-1, p. 40)

31 FLOWERS   Danté is making bunches of flowers for his mother, his three aunts, and his grandfather. Each bunch will have 10 flowers. How many flowers will Danté need to make 5 bunches?

_____

Draw an array to model each expression.

1 $7 \times 3 =$ _____

2 $8 \times 6 =$ _____

Rewrite the factors in distributive form. Find each product.

3 $5 \times 12 = ($_____$\times 10) + ($_____$\times 2)$

$=$ _____ $+$ _____

$=$ _____

4 $12 \times 7 = ($_____$\times 7) + ($_____$\times 7)$

$=$ _____ $+$ _____

$=$ _____

5 $11 \times 4 = ($_____$\times 4) + ($_____$\times 4)$

$=$ _____ $+$ _____

$=$ _____

6 $11 \times 8 = (10 \times$_____$) + (1 \times$_____$)$

$=$ _____ $+$ _____

$=$ _____

Find each product. Show your work.

7 $6 \times 11 =$ _____

8 $8 \times 9 =$ _____

9 $7 \times 4 =$ _____

10 $11 \times 2 =$ _____

11 $8 \times 12 =$ _____

12 $5 \times 9 =$ _____

Solve.

13 **ZOO** On Saturday 12 school groups visited the zoo. Each group had 20 students. How many students visited the zoo on Saturday?

_____

14 **SUPPLIES** Oscar bought 9 packs of pencils. There are 12 pencils in each pack. How many pencils did Oscar buy?

_____

Lesson 2-5 Multiply Greater Numbers

KEY Concept

Traditional Multiplication Method

$$\begin{array}{r} 1 \\ 43 \\ \times\ 35 \\ \hline 215 \\ +\ 1\,290 \\ \hline 1{,}505 \end{array}$$

Partial Products Method

$$\begin{array}{r} 43 \\ \times\ 35 \\ \hline 15 \\ 200 \\ 90 \\ +\ 1{,}200 \\ \hline 1{,}505 \end{array}$$

$5 \times 3 = 15$
$5 \times 40 = 200$
$30 \times 3 = 90$
$30 \times 40 = 1{,}200$

estimate
a number close to an exact value; an estimate indicates *about* how much

partial products method
a way to multiply; the value of each digit in one factor is multiplied by the value of each digit in the other factor; the product is the sum of its partial products

Before you multiply, you should **estimate** your answer. Then check your actual answer for reasonableness.

Example 1

Find the product of 36 and 82. Use the partial products method.

1. Estimate. **40 × 80 = 3,200**

2. Rewrite the problem in a vertical format.

3. Multiply 2 times the ones column.
   $2 \times 6 = 12$

4. Multiply 2 times the tens column.
   $2 \times 30 = 60$

5. Multiply 80 times the ones column.
   $80 \times 6 = 480$

6. Multiply 80 times the tens column.
   $80 \times 30 = 2{,}400$

$$\begin{array}{r} 36 \\ \times\ 82 \\ \hline 12 \\ 60 \\ 480 \\ +2{,}400 \\ \hline 2{,}952 \end{array}$$

7. Add the partial products.
   $12 + 60 + 480 + 2{,}400 = 2{,}952$

8. $36 \times 82 = 2{,}952$

YOUR TURN!

Find the product of 28 and 57. Use the partial products method.

1. Estimate. $30 \times 60 = $ _____

2. Rewrite the problem in a vertical format.

3. Multiply _____ times the ones column.
   $7 \times 8 = $ _____

4. Multiply _____ times the tens column.
   $7 \times 20 = $ _____

5. Multiply 50 times the ones column.
   $50 \times 8 = $ _____

6. Multiply 50 times the tens column.
   $50 \times 20 = $ _____

$$\begin{array}{r} 28 \\ \times\ 57 \\ \hline \\ \hline \end{array}$$

7. Add the partial products.
   _____ + _____ + _____ + _____
   = _____

8. $28 \times 57 = $ _____

Copyright © by The McGraw-Hill Companies, Inc.

68   **Chapter 2** Multiplication

Example 2

Find the product of 76 and 14. Use the traditional multiplication method.

1. Estimate.
   80 × 10 = 800

2. Rewrite the problem in a vertical format.

3. Multiply 4 times the digit in the ones column.
   4 × 6 = 24
   Write the tens digit above the tens column. Write the ones digit in the product under the ones column.

   $$\begin{array}{r} 2 \\ 76 \\ \times\ 14 \\ \hline 4 \end{array}$$

4. Multiply 4 times the tens column.
   4 × 7 = 28
   Add the two tens to get 30.

   $$\begin{array}{r} 2 \\ 76 \\ \times\ 14 \\ \hline 304 \end{array}$$

5. Multiply the value of the digit in the tens place times 6.
   10 × 6 = 60

   $$\begin{array}{r} 76 \\ \times\ 14 \\ \hline 304 \\ 60 \end{array}$$

6. Multiply the value in the tens column by 10.
   10 × 70 = 700
   Write the 7 in the hundreds place.

   $$\begin{array}{r} 76 \\ \times\ 14 \\ \hline 304 \\ +\ 760 \\ \hline 1{,}064 \end{array}$$

7. Find the sum of the two products.

8. 76 × 14 = 1,064

   Compare to your estimate for reasonableness.

YOUR TURN!

Find the product of 56 and 32. Use the traditional multiplication method.

1. Estimate.
   _____ × _____ = _____

2. Rewrite the problem in a vertical format.

3. Multiply _____ times the digit in the ones column.

   $$\begin{array}{r} 56 \\ \times\ 32 \end{array}$$

   2 × 6 = _____

4. Multiply _____ times the tens column.

   _____

   2 × 5 = _____

   Add the _____ ten(s) to get _____.

5. Multiply each place value by the tens digit.

   30 × 6 = _____

6. Multiply the value in the tens column by 30.

   30 × 50 = _____

   Add _____ hundred(s) for _____.

7. Find the sum of the two products.

8. 56 × 32 = _____

   Compare to your estimate for reasonableness.

GO ON

Who is Correct?

Find the product of 26 and 47.

Vivian	Alonso	Lamar
26	26	26(40+7)
× 47	× 47	= (26 × 40)+(26 × 7)
42	42	= 1,040+182
14	140	= 1,222
24	240	
+ 8	+ 800	
88	1,222	

Circle correct answer(s). Cross out incorrect answer(s).

 Guided Practice

Estimate each product.

1 95 × 47

Round each factor to the greatest place value.
Find the estimated product.

_____ × _____ = _____

2 70 × 32

Round each factor to the greatest place value.
Find the estimated product.

_____ × _____ = _____

Step by Step Practice

3 Find the product of 14 and 89. Use the traditional multiplication method.

Step 1 Rewrite the problem in a vertical format.

Step 2 Multiply each place value by the ones digit.

Step 3 Multiply each place value by the tens digit.

Step 4 Find the sum of the two products.

Step 5 14 × 89 = _____

$$\begin{array}{r} 14 \\ \times\ 89 \\ \hline \end{array}$$

Find each product. Use the traditional multiplication method.

4 $11 \times 71 =$ _____

$$\begin{array}{r} 11 \\ \times\ 71 \\ \hline \end{array}$$

5 $12 \times 101 =$ _____

Find each product. Use the partial products method.

6 $39 \times 83 =$ _____

7 $28 \times 94 =$ _____

Step by Step Problem-Solving Practice

Solve.

8 **SCHOOL**   The teacher's aide is grading tests. Each test has 45 questions. Each student took 4 tests. If there are 28 students in the class, how many test questions are there in all?

Problem-Solving Strategies	
☐ Draw a diagram.	
☑ Use logical reasoning.	
☐ Solve a simpler problem.	
☐ Work backward.	
☐ Make a table.	

Understand   Read the problem. Write what you know.

Each student took _____ tests.

Each test has _____ questions.

There are _____ students in the class.

Plan   Pick a strategy. One strategy is to use logical reasoning. Find the number of questions on 4 tests. Then multiply by the number of students to find the number of test questions in all.

Solve   number of questions × number of tests   = questions on 4 tests
questions on 4 tests   × number of students = test questions in all

Use the values from the problem to solve.

_____ × _____ = _____

_____ × _____ = _____

There are _____ test questions in all.

Check   Look at your solution. Did you answer the question?

GO ON

9 **BUSINESS** Ginny is digging for clams to sell to a local seafood store. For every shovel of sand she digs up, she finds 16 small clams. If she digs 27 shovels of sand, how many clams could she expect to find?
Check off each step.

_____ Understand: I circled key words.

_____ Plan: To solve the problem, I will _____.

_____ Solve: The answer is _____.

_____ Check: I checked my answer by _____.

10 **COOKING** When making meat loaf, Ruben uses 12 ounces of breadcrumbs. His mother uses the same recipe at her restaurant, but she multiplies the recipe to make 32 meat loaves. How many ounces of breadcrumbs will his mother use?

_____

11 **Reflect** Compare two different ways of multiplying 13×17.

_____

_____

 ## Skills, Concepts, and Problem Solving

Find each product. Use the traditional multiplication method.

12 $45 \times 87 =$ _____

13 $56 \times 91 =$ _____

14 $11 \times 15 =$ _____

15 $13 \times 14 =$ _____

16 $28 \times 51 =$ _____

17 $32 \times 60 =$ _____

Find each product. Use the partial products method.

18 $39 \times 63 =$ _____

19 $18 \times 22 =$ _____

20 $32 \times 16 =$ _____

21 $32 \times 31 =$ _____

22 $42 \times 14 =$ _____

23 $58 \times 22 =$ _____

Find each product. Show your work.

24 $19 \times 19 =$ _____

25 $29 \times 21 =$ _____

26 $48 \times 77 =$ _____

27 $25 \times 87 =$ _____

28 $26 \times 31 =$ _____

29 $44 \times 62 =$ _____

30 $17 \times 45 =$ _____

31 $51 \times 23 =$ _____

Solve

32 COOKING   Ricky is making sandwiches for a banquet. Each package of cheese has 16 slices. He uses 14 packages. How many slices of cheese did he use in all?

_____

33 PHOTOGRAPHY   Tessa's online photo album holds 50 pictures. She has 18 photo albums. How many pictures does she have in all?

_____

GO ON

Solve.

34 MUSIC Look at the photo at the right. If Martino has 63 CDs, how many total songs does Martino have?

_____

MUSIC Each CD has 11 songs.

35 FITNESS Lawana runs 3 miles every day. Michelle runs 4 miles a day on weekdays. Michelle does not run on the weekends. After 45 weeks, who will have run more? By how much?

_____

Vocabulary Check **Write the vocabulary word that completes each sentence.**

36 A number close to an exact value is called a(n) _____.

37 Writing in Math Sophie multiplied 45 × 18. What mistake did she make?

$$\begin{array}{r} \overset{4}{45} \\ \times\, 18 \\ \hline 360 \\ +345 \\ \hline 705 \end{array}$$

_____

_____

▶ Spiral Review

Solve. (Lesson 2-3, p. 55)

38 GROCERY There are 7 egg cartons on the shelf at the corner market. Each egg carton contains 1 dozen eggs. What is the total number of eggs on the shelf?

_____

Find each product. (Lesson 2-4, p. 61)

39 $11 \times 8 =$ _____

40 $12 \times 7 =$ _____

41 $12 \times 13 =$ _____

42 $11 \times 46 =$ _____

43 $12 \times 31 =$ _____

44 $11 \times 52 =$ _____

STOP

Multiplication and Division

KEY Concept

Multiplication and **division** are opposite, or **inverse operations**. They undo each other.

There are 4 cartons of eggs. Each carton has 12 eggs. There are 48 eggs in all.

$$4 \quad \times \quad 12 \quad = \quad 48$$
factor   factor   product

There are 48 eggs in all. Each carton has 12 eggs. There are 4 cartons.

$$48 \quad \div \quad 12 \quad = \quad 4$$
product   factor   factor

> In multiplication, the product is the missing number. In division, one of the factors is the missing number.

A **fact family** is a group of related facts using the same numbers.

$$4 \quad \times \quad 12 \quad = \quad 48$$
$$12 \quad \times \quad 4 \quad = \quad 48$$
$$48 \quad \div \quad 12 \quad = \quad 4$$
$$48 \quad \div \quad 4 \quad = \quad 12$$

VOCABULARY

division
an operation on two numbers in which the first number is split into the same number of equal groups as the second number
Example: 6 ÷ 3 means 6 is divided into 3 groups of equal size.

fact family
a group of related facts using the same numbers
Example: 5 × 3 = 15;
3 × 5 = 15;
15 ÷ 3 = 5; 15 ÷ 5 = 3

inverse operations
operations that undo each other
Example: Multiplication and division are inverse operations.

multiplication
an operation on two numbers to find their product; it can be thought of as repeated addition

You can use what you know about multiplication to help you learn more about division.

GO ON

Example 1

Write the fact family for the array.

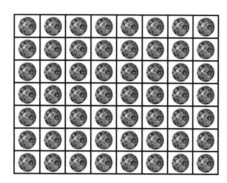

1. Write two multiplication sentences shown by the array.

   $7 \times 8 = 56$ $8 \times 7 = 56$

2. Write two division sentences using the same numbers.

   $56 \div 7 = 8$ $56 \div 8 = 7$

YOUR TURN!

Write the fact family for the array.

1. Write two multiplication sentences shown by the array.

   _____   _____

2. Write two division sentences using the same numbers.

   _____   _____

Who is Correct?

Write a division sentence related to $45 \times 9 = 405$.

Farris	Nadina	Makalla
$405 \div 45 = 9$	$405 \div 9 = 40$	$9 \times 45 = 405$

Circle correct answer(s). Cross out incorrect answer(s).

▶ Guided Practice

Write a number family for the array.

1 Write two multiplication sentences shown by the array.

   _____

   _____

2 Write two division sentences with the same numbers.

   _____

   _____

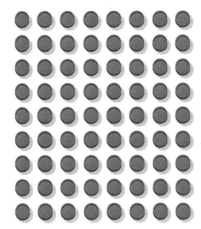

Step by Step Practice

3 Complete each equation. Write the numbers in the fact family.

   13 × 20 = ____

   20 × ____ = 260

   260 ÷ ____ = 13

   260 ÷ ____ = 20

   ____ , ____ , ____

> HINT: Each equation is missing one number.
> · Find the missing number in one equation.
> · Use your answer to complete the remaining equations.

Step 1 Find the third number in the first equation. Find the product of 13 and 20.

Step 2 What is the missing factor in the second equation?

Step 3 Which number completes the third equation? ____

Step 4 Which number completes the fourth equation? ____

Step 5 Write the numbers in the fact family.

Step 6 Skip count by 20s to check.

| 1 | 2 | 3 | 4 | 5 | 6 | 7 | 8 | 9 | 10 | 11 | 12 | 13 |

____ , ____ , ____ , ____ , ____ , ____ , ____ , ____ , ____ , ____ , ____ , ____ , ____

GO ON

Complete each equation using the numbers in the fact family.

4 $8 \times 17 = \rule{1cm}{0.15mm}$

$17 \times \rule{1cm}{0.15mm} = 136$

$136 \div 8 = \rule{1cm}{0.15mm}$

$136 \div \rule{1cm}{0.15mm} = 8$

5 $36 \times 5 = \rule{1cm}{0.15mm}$

$5 \times \rule{1cm}{0.15mm} = 180$

$180 \div 5 = \rule{1cm}{0.15mm}$

$180 \div \rule{1cm}{0.15mm} = 5$

6 $24 \times 19 = \rule{1cm}{0.15mm}$

$19 \times \rule{1cm}{0.15mm} = 456$

$456 \div 24 = \rule{1cm}{0.15mm}$

$456 \div \rule{1cm}{0.15mm} = 24$

Step by Step Problem-Solving Practice

Problem-Solving Strategies

☑ Draw a model.
☐ Use logical reasoning.
☐ Make a table.
☐ Solve a simpler problem.
☐ Work backward.

7 SCHOOL   The auditorium chairs are set up for a school performance. There are 4 sections, each with 128 chairs. How many chairs are set up in the auditorium?

Understand   Read the problem. Write what you know.

There are _____ sections of chairs.

Each section has _____ chairs.

Plan   Pick a strategy. One strategy is to draw a model. Draw 4 sections. Show that each section contains 128 chairs.

Solve   There are 4 sections. Each section has 128 chairs.

| 128 chairs | 128 chairs | 128 chairs | 128 chairs |

Multiply 128 by 4 to find the total number of chairs.

$$\begin{array}{r} 128 \\ \times\ 4 \\ \hline \end{array}$$

There are _____ chairs set up in the auditorium.

Check   Use a related multiplication or division equation to check your work.

8 MAP   Jin is making a map of Junction City. There are 6 neighborhoods on the map. Each neighborhood has 145 houses. How many houses will be shown on Jin's map? Check off each step.

_____ **Understand: I circled key words.**

_____ **Plan: To solve this problem, I will** _____.

_____ **Solve: The answer is** _____.

_____ **Check: I checked my answer by** _____.

9 CAMPING   72 children are spending a week at camp. There are 12 cabins at the camp. An equal number of campers will sleep in each cabin. How many campers will sleep in each cabin?

_____

10 Reflect   What does it mean to say that multiplication and division are _inverse operations_? Give an example.

_____

_____

_____

▶ Skills, Concepts, and Problem Solving

Write a number family for the array.

11 Write two multiplication sentences.

_____

_____

Write two division sentences.

_____   _____

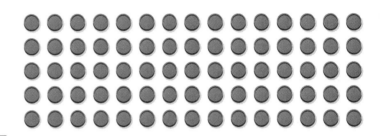

Complete each equation. Write the numbers in the fact family.

12 $6 \times 45 =$ _____

$45 \times$ _____ $= 270$

$270 \div 45 =$ _____

$270 \div$ _____ $= 45$

The numbers in the fact family are:

_____, _____, _____.

13 $13 \times 7 =$ _____

$7 \times$ _____ $= 91$

$91 \div 7 =$ _____

$91 \div$ _____ $= 7$

The numbers in the fact family are:

_____, _____, _____.

Fill in the blanks to make each equation correct.

14 $21 \times$ _____ $= 903$

15 $9 \times$ _____ $= 450$

16 $88 \times$ _____ $= 616$

17 $888 \div 4 =$ _____

18 $45 \div 3 =$ _____

19 $132 \div 2 =$ _____

GO ON

Use either multiplication or division to solve each problem. Use a related multiplication or division equation to check your work.

20 PAINT   It takes 5 gallons of paint to paint each room in a school. The school has 26 rooms. How many gallons of paint will be needed to paint all the rooms? _____

Check: _____

21 LIBRARY   The public library has 900 new books. There are 20 empty shelves where the books will be displayed. If the same number of books must be put on each shelf, how many books will go on each shelf? _____

Check: _____

Vocabulary Check   **Write the vocabulary word that completes each sentence.**

22 A group of related facts using the same numbers is called a(n) _____.

23 Writing in Math   Describe an everyday situation where you would use division. Give an example.

_____

_____

 Spiral Review

Find each product. Show your work. (Lesson 2-5, p. 68)

24 14 × 99 _____

25 62 × 23 _____

Solve. (Lesson 2-4, p. 61)

26 HORSES   Mr. Ortega is putting horseshoes on his 12 horses. Each horse needs 4 horseshoes. How many horseshoes will Mr. Ortega need for all of his horses?

_____

Find each product. Use the partial products method.

1 27	**2** 46
× 16	× 32

      _____ = 6 × 7   _____ = 2 × 6

      _____ = 6 × 20   _____ = 2 × 40

      _____ = 10 × 7   _____ = 30 × 6

+ _____ = 10 × 20   + _____ = 30 × 40

Find each product. Use the traditional multiplication method.

3 56	**4** 64
× 17	× 23

Complete each equation. Write the numbers in the fact family.

5   15 × 27 = _____   **6**   49 × _____ = 539

     27 × _____ = 405   _____ × 49 = 539

     405 ÷ _____ = 15   539 ÷ _____ = 49

     405 ÷ 15 = _____   539 ÷ 49 = _____

     _____, _____, _____   _____, _____, _____

Solve.

7   **BUSES**   Ridgewood Middle School uses buses to take fifth graders on a field trip. One bus can hold 40 students. How many buses will be needed to take 320 fifth graders on a field trip?

_____

8   **BEADS**   Tammy wants to make a bracelet for all 18 girls on her softball team. It takes 12 beads to make a bracelet. How many beads does Tammy need?

_____

Vocabulary and Concept Check

array, *p. 47*

Commutative Property
of Multiplication, *p. 47*

Distributive Property
of Multiplication, *p. 61*

division, *p. 75*

estimate, *p. 68*

fact family, *p. 75*

factor, *p. 40*

Identity Property of
Multiplication, *p. 40*

inverse operations, *p. 75*

multiple, *p. 55*

multiplication, *p. 40*

partial products method, *p. 68*

pattern, *p. 61*

product, *p. 40*

Zero Property of Multiplication, *p. 40*

Write the vocabulary word that completes each sentence.

1 A number that is multiplied by another number to find a product is called a(n) _____.

2 The _____ states that any number multiplied by zero is zero.

3 _____ are operations that undo each other.

4 A number close to an exact value is a(n)

_____.

5 A group of related facts using the same numbers is called a(n)

_____.

Label each diagram below. Write the correct vocabulary term in each blank.

6 _____

$$4: 4, 8, 12, 16, 20, 24$$

7 _____

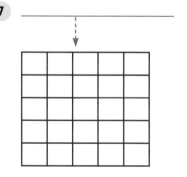

Lesson Review

2-1 Multiply by 0, 1, 5, and 10 (pp. 40–46)

Find each product.

8 $16 \times 0 =$ _____

9 $92 \times 10 =$ _____

10 $5 \times 15 =$ _____

11 $1 \times 310 =$ _____

12 $0 \times 402 =$ _____

13 $71 \times 10 =$ _____

14 $48 \times 1 =$ _____

15 $20 \times 5 =$ _____

Example 1

Find the product of 5 and 13.

1. Rewrite the problem in a vertical format.

2. Multiply the number in the ones column by 5.
   $3 \times 5 = 15$
   Write the tens digit above the tens column. Write the ones digit under the ones column.

$$\begin{array}{r} {}^{1}13 \\ \times\ 5 \\ \hline 5 \end{array}$$

3. Multiply 5 times the digit in the tens column.
   $5 \times 1 = 5$
   Add one regrouped tens for a total of 6 tens.

$$\begin{array}{r} {}^{1}13 \\ \times\ 5 \\ \hline 65 \end{array}$$

4. $5 \times 13 = 65$

2-2 Multiply by 2, 3, 4, and 6 (pp. 47–53)

Find each product.

16 $2 \times 28 =$ _____

17 $3 \times 17 =$ _____

18 $51 \times 4 =$ _____

19 $36 \times 6 =$ _____

20 $17 \times 2 =$ _____

21 $12 \times 6 =$ _____

22 $4 \times 30 =$ _____

23 $21 \times 3 =$ _____

Example 2

Find the product of 4 and 32.

1. Rewrite the problem in a vertical format.

2. Multiply the number in the ones column by 4. $4 \times 2 = 8$
   Write the ones digit under the ones column.

$$\begin{array}{r} 32 \\ \times\ 4 \\ \hline 8 \end{array}$$

3. Multiply 4 times the digit in the tens column.
   $4 \times 3 = 12$
   Write the tens digits under the tens column.

$$\begin{array}{r} 32 \\ \times\ 4 \\ \hline 128 \end{array}$$

4. $4 \times 32 = 128$

2-3 Multiply by 7, 8, and 9 (pp. 55–60)

Find each product.

24 $7 \times 6 =$ _____

25 $8 \times 8 =$ _____

26 $9 \times 7 =$ _____

27 $31 \times 8 =$ _____

28 $56 \times 9 =$ _____

29 $19 \times 7 =$ _____

30 $8 \times 15 =$ _____

31 $9 \times 12 =$ _____

Example 3

Use a pattern to find the product of 5×8.

1. Skip count by 8s to write the multiples of 8.

2. Multiples of 8: 8, 16, 24, 32, 40, 48, 56, 64

$1 \times 8 = 8$	$2 \times 8 = 16$	$3 \times 8 = 24$
$4 \times 8 = 32$	$5 \times 8 = 40$	$6 \times 8 = 48$
$7 \times 8 = 56$	$8 \times 8 = 64$	$9 \times 8 = 72$
$10 \times 8 = 80$		

3. $5 \times 8 = 40$

2-4 Multiply by 11 and 12 (pp. 61–66)

Find each product.

32 $4 \times 11 =$ _____

33 $12 \times 6 =$ _____

34 $11 \times 12 =$ _____

35 $3 \times 12 =$ _____

36 $11 \times 8 =$ _____

37 $12 \times 7 =$ _____

38 $3 \times 11 =$ _____

39 $9 \times 12 =$ _____

Example 4

Use an array model to find 12×8.

1. Rewrite 12×8 in distributive form.
   $(10 \times 8) + (2 \times 8)$

2. Draw an array using two sets of counters.

   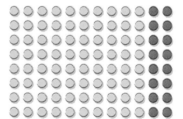

3. The array shows $10 \times 8 = 80$ and $2 \times 8 = 16$. So, $80 + 16 = 96$.

4. $12 \times 8 = 96$

2-5 Multiply Greater Numbers (pp. 68–73)

Find each product.

40   13
      × 18

41   47
      × 62

42   26
      × 53

43   71
      × 29

Example 5

Find the product of 31 and 42. Use the partial products method.

1. Rewrite the problem in a vertical format.

2. Multiply 2 times the ones column.
   $2 \times 1 = 2$

3. Multiply 2 times the tens column.
   $2 \times 30 = 60$

4. Multiply 40 times the ones column.
   $40 \times 1 = 40$

5. Multiply 40 times the tens column.
   $40 \times 30 = 1{,}200$

6. Add the partial products.
   $2 + 60 + 40 + 1{,}200 = 1{,}302$

7. $31 \times 42 = 1{,}302$.

```
      31
    × 42
       2
      60
      40
   1,200
   1,302
```

2-6 Multiplication and Division (pp. 75–80)

Complete each equation. Write the numbers in the number family.

44 $18 \times 62 = \underline{\hspace{1cm}}$

   $62 \times \underline{\hspace{1cm}} = 1{,}116$

   $1{,}116 \div \underline{\hspace{1cm}} = 62$

   $1{,}116 \div \underline{\hspace{1cm}} = 18$

   $\underline{\hspace{1cm}}, \underline{\hspace{1cm}}, \underline{\hspace{1cm}}$

45 $45 \times \underline{\hspace{1cm}} = 450$

   $\underline{\hspace{1cm}} \times 45 = 450$

   $450 \div 10 = \underline{\hspace{1cm}}$

   $450 \div \underline{\hspace{1cm}} = 10$

   $\underline{\hspace{1cm}}, \underline{\hspace{1cm}}, \underline{\hspace{1cm}}$

Example 6

Write the fact family for the array.

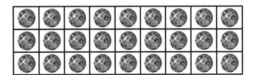

1. Write two multiplication sentences shown by the array.
   $9 \times 3 = 27$
   $3 \times 9 = 27$

2. Write two division sentences with the same numbers.
   $27 \div 9 = 3$
   $27 \div 3 = 9$

Chapter Test

Draw an array to model each expression.

1   $9 \times 3 = $ _____

2   $8 \times 8 = $ _____

Find each product. Show your work.

3   $0 \times 481 = $ _____

4   $97 \times 1 = $ _____

5   $7 \times 5 = $ _____

6   $18 \times 10 = $ _____

7   $13 \times 2 = $ _____

8   $3 \times 18 = $ _____

9   $21 \times 4 = $ _____

10   $6 \times 14 = $ _____

11   $16 \times 7 = $ _____

12   $28 \times 8 = $ _____

13   $9 \times 20 = $ _____

14   $11 \times 6 = $ _____

15   $12 \times 4 = $ _____

16   $17 \times 5 = $ _____

17   $12 \times 8 = $ _____

GO ON

Use partial products to find the product.

18   29
       × 38

19   44
       × 57

Solve.

20 **FRUIT** Brooke is picking strawberries from her garden. She picked 9 pints and 15 fit into a pint. How many strawberries will Brooke have?

_____

21 **FITNESS** Ricco is lifting weights at the gym. Ricco has 4 weights on the barbell he is bench pressing. Each weight is 20 pounds. How much can Ricco bench press?

_____

Correct the mistakes.

22 **SNACKS** Charles wanted to share his box of raisins with Mila and Miles. He counted 36 raisins altogether. He said, "We can each have 14 raisins." Is Charles' statement correct? How do you know?

_____

_____

STOP

Choose the best answer and fill in the corresponding circle on the sheet at right.

1 $36 \times 1 = 36$ is an example of which property?

    A Distributive Property

    B Zero Property of Multiplication

    C Identity Property of Multiplication

    D Commutative Property of Multiplication

2 $68 \times 2 =$

    A 70   **C** 163

    B 136   **D** 34

3 Find the product of 12 and 52.

    A 624   **C** 246

    B 64   **D** 46

4 Which is an example of Commutative Property of Multiplication?

    A $6(4 + 1) = (6 \times 4) + (6 \times 1)$
        $= 24 + 6 = 30$

    B $6 \times 0 = 0$

    C $6 \times 1 = 6$

    D $6 \times 4 = 24$ and $4 \times 6 = 24$

5 The first four multiples of 9 are...

    A 1, 3, 3, 9   **C** 36

    B 9, 18, 27, 36   **D** 9, 10, 11, 12

6 Grace cut a rope into 7 pieces. Each piece is 13 ft long. How long was the original rope?

    A 21 feet   **C** 98 feet

    B 91 feet   **D** 104 feet

7 Which expression does this array model?

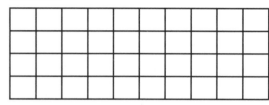

    A $10 \div 4$ **C** 4×10

    B 40   **D** 40×4

8 Bruno is walking dogs for his dog walking business. If he counts 36 dog legs, how many dogs is he walking?

    A 9 dogs   **C** 4 dogs

    B 2 dogs   **D** 12 dogs

9 $86 \times 7 =$

    A 93   **C** 206

    B 522   **D** 602

GO ON

10 Which is an example of an inverse operation?

    A addition and multiplication

    B multiplication and division

    C subtraction and division

    D addition and division

11 The cafeteria has 5 long tables. Each table can seat 10 students. How many students can sit at all of the tables?

    A 2

    B 20

    C 50

    D 55

12 The product of 77 and 7 is _____.

    A 84

    B 70

    C 11

    D 539

ANSWER SHEET

Directions: Fill in the circle of each correct answer.

1 (A) (B) (C) (D)
2 (A) (B) (C) (D)
3 (A) (B) (C) (D)
4 (A) (B) (C) (D)
5 (A) (B) (C) (D)
6 (A) (B) (C) (D)
7 (A) (B) (C) (D)
8 (A) (B) (C) (D)
9 (A) (B) (C) (D)
10 (A) (B) (C) (D)
11 (A) (B) (C) (D)
12 (A) (B) (C) (D)

Success Strategy

Double check your answers after you finish. Read each problem and all of the answer choices. Put your finger on each bubble you filled in to make sure it matches the answer for each problem.

STOP

Division

Let's play a game of soccer.

Division helps us make up the teams. If there are 18 people who want to play, you would divide 18 by 2 to get 2 teams of 9 people.

STEP **1** Quiz

Are you ready for Chapter 3? Take the Online Readiness Quiz at *glencoe.com* to find out.

STEP **2** Preview

Get ready for Chapter 3. Review these skills and compare them with what you'll learn in this chapter.

What You Know	What You Will Learn

What You Know

You know how to subtract.

Examples: $8 - 2 = 6$
$6 - 2 = 4$
$4 - 2 = 2$
$2 - 2 = 0$

TRY IT!

1   $32 - 8 = $ _____

2   $40 - 10 = $ _____

3   $16 - 4 = $ _____

4   $20 - 1 = $ _____

What You Will Learn

Lesson 3-1

Division is repeated subtraction.

$8 \div 2 = ?$

$8 - 2 = 6$ Subtract 2 one time.
$6 - 2 = 4$ Subtract 2 two times.
$4 - 2 = 2$ Subtract 2 three times.
$2 - 2 = 0$ Subtract 2 four times.

So, $8 \div 2 = 4$.

You know that addition and subtraction are inverse operations.

Example: $5 + 7 = 12$
$12 - 7 = 5$

TRY IT!

Rewrite each equation using an inverse operation.

5   $9 - 5 = 4$

_____

6   $15 + 10 = 25$

_____

7   $64 - 31 = 33$

_____

8   $22 + 76 = 98$

_____

Lessons 3-2 through 3-8

Multiplication and division are **inverse operations**.

$5 \times 7 = 35$

$35 \div 7 = 5$

$9 \times 3 = 27$

$27 \div 3 = 9$

Lesson 3-1 Model Division

KEY Concept

Division is the **inverse operation** of multiplication. You use multiplication facts when you divide. Division is used to make groups of equal size.

If you had eight pretzels and wanted to share them with a friend, you would divide the pretzels into two groups.

The division can be written three ways.

horizontal method	vertical method	fraction method
$8 \div 2 = 4$	$2\overline{)8}$ with 4 above	$\dfrac{8}{2} = 4$

- The **dividend** in these problems is 8.
- The **divisor** in these problems is 2.
- The **quotient** in these problems is 4.

Arrays can help you find the quotient of a division problem.

$8 \div 2 = 4$

1	3	5	7
2	4	6	8

} number of rows = divisor

number of columns = quotient

number in all		number of rows		number of columns
8	$\div$	2	$=$	4
dividend		**divisor**		**quotient**

Use multiplication to check your division.

$8 \div 2 = 4$ is correct because $2 \times 4 = 8$.

VOCABULARY

array
objects or symbols displayed in rows of the same length and columns of the same length; the length of a row might be different from the length of a column

dividend
the number that is being divided

division
to separate into equal groups
Example: $6 \div 3$ means 6 is divided into 3 groups of equal size.

divisor
the number by which the dividend is being divided

inverse operations
operations that undo each other
Example: Multiplication and division are inverse operations.

quotient
the answer or result of a division problem

Think of fact families when dividing. A fact family has two multiplication sentences and two division sentences.

$$4 \times 2 = 8 \qquad 8 \div 2 = 4$$
$$2 \times 4 = 8 \qquad 8 \div 4 = 2$$

Example 1

Draw an array to model the expression 6 ÷ 3.

1. Identify the dividend (the first number). This is the *total number of rectangles* in the array.   **6**

2. Identify the divisor (the second number). This represents the number of *rows*.   **3**

3. Draw an array with 6 rectangles in 3 rows.

1	4
2	5
3	6

2 columns = quotient

4. The number of columns in the array is the quotient.   **2**

5. Check by multiplying the quotient by the divisor. The product should be the dividend.   **3 × 2 = 6**

YOUR TURN!

Draw an array to model the expression 12 ÷ 4.

1. Identify the dividend. _____ This is the _____ number of rectangles in the array.

2. Identify the divisor. _____ This represents the number of _____.

3. Draw an array with _____ rectangles in _____ rows.

1	5	9
2	6	10
3	7	11
4	8	12

_____ columns = quotient

4. The number of columns in the array is the quotient.

_____

5. Check by multiplying the quotient by the divisor.
3 × _____ = 12

Example 2

Draw a model of 10 ÷ 2 using circles and tally marks.

1. What is the divisor?   **2**

   Draw 2 circles.

2. What is the dividend?   **10**

   Use tally marks to divide the 10 into 2 groups.

   Place a tally mark in each circle as you count until you have drawn 10 tally marks.

3. How many tally marks are in each circle?   **5**

   Write the problem with the quotient.   **10 ÷ 2 = 5**

4. Check.   **2 × 5 = 10 ✓**

GO ON

Draw a model of 15 ÷ 3 using circles and tally marks.

1. What is the divisor? _____ Draw _____ circles.

2. What is the dividend? _____

   Use tally marks to divide the _____ into _____ groups.

   Place a tally mark in each circle as you count until you have drawn _____ tally marks.

3. How many tally marks are in each circle? _____

   Write the problem with the quotient. 15 ÷ 3 = _____

4. Check. 3 × _____ = 15

Example 3

Write 6 ÷ 3 in two different formats.

1. What number is the divisor? 3
   The divisor goes in front of the division bracket.

2. Write the vertical format. $3\overline{)6}$
   Read as, "Six divided by three."

3. Write as a fraction. $\dfrac{6}{3}$
   Read as, "Six divided by three."

Write 15 ÷ 5 in two different formats.

1. What number is the divisor? _____

2. Write the vertical format. _____
   Read as,

   "_____ divided by _____."

3. Write as a fraction. _____
   Read as,

   "_____ divided by _____."

Example 4

Write the division facts from the fact family of 5 × 2 = 10.

1. Write the division fact using the first factor as the divisor.
   10 ÷ 5 = 2

2. Write another division fact using the second factor as the divisor.
   10 ÷ 2 = 5

Write the division facts from the fact family of 7 × 2 = 14.

1. Write the division fact using the first factor as the divisor.

   _____ ÷ _____ = _____

2. Write another division fact using the second factor as the divisor.

   _____ ÷ _____ = _____

Who is Correct?

Write 16 ÷ 4 in two different formats.

Paulita

$16\overline{)4}$ and $\frac{16}{4}$

Zola

$4\overline{)16}$ and $\frac{16}{4}$

Augustin

$4\overline{)16}$ and $\frac{4}{16}$

Circle correct answer(s). Cross out incorrect answer(s).

▶ Guided Practice

1 Draw an array to model the expression 12 ÷ 3.

2 Draw a model of 12 ÷ 2 using circles and tally marks.

Step by Step Practice

3 Draw an array to model the expression 12 ÷ 6.

Step 1 Identify the dividend. _____

This is the total number of rectangles in the array.

Create an array that has _____ rectangles in

_____ rows.

Step 2 Identify the divisor. _____

This represents the number of rows.

Step 3 The number of columns in the array is the quotient.

_____

Step 4 Check your division by multiplying the quotient by the divisor.

_____.

GO ON

Draw an array to model each expression.

4 9 ÷ 3   How many rows? _____
           How many rectangles? _____
           How many columns? _____

5 8 ÷ 4

6 14 ÷ 7

7 6 ÷ 3

8 16 ÷ 2

Step by Step Problem-Solving Practice

Solve.

9 **GARDENS**   Berto is planting a garden with 18 seeds. He wants to have 3 rows of seeds with the same number of seeds in each row. How many seeds will be in each row?

Problem-Solving Strategies
- ☑ Draw a diagram.
- ☐ Use logical reasoning.
- ☐ Solve a simpler problem.
- ☐ Work backward.
- ☐ Guess and check.

Understand   Read the problem. Write what you know.

There are _____ seeds. There will be _____ rows.

Plan   Pick a strategy. One strategy is to draw a diagram.

Solve   Make an array with the correct number of rows.

Continue making columns until there are _____ rectangles.

The diagram shows how the seeds will be planted.

There are _____ rows and _____ columns.
The number of columns represents the _____.
There will be _____ seeds in each row.

Check   Think about the fact family. Is the division fact part of the family?

10 **PROJECT**   Vito and 3 of his classmates are to study the 12 animals from the Chinese calendar. How many animals will each of them have to study? Check off each step.

_____ Understand: I circled key words.

_____ Plan: To solve the problem, I will _____.

_____ Solve: The answer is _____.

_____ Check: I checked my answer by _____.

PROJECT
Chinese calendar

11 **ENTERTAINMENT**   José is arranging 20 chairs for a party. He wants to have 5 rows of chairs with the same number of chairs in each row. How many chairs will he put in each row?

_____

12 **Reflect**   Explain how multiplication is the inverse operation for division.

_____

_____

_____

▶ Skills, Concepts, and Problem Solving

Draw an array to model each expression.

13 20 ÷ 5

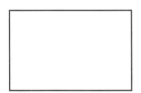

14 8 ÷ 4

Draw a model using circles and tally marks for each expression.

15 18 ÷ 3

16 9 ÷ 3

17 3 ÷ 1

18 8 ÷ 2

GO ON

Write each expression in two different formats.

19 $25 \div 5$ _____ _____

20 $16 \div 8$ _____ _____

Write the division facts from each fact family.

21 $6 \times 3 = 18$ _____ **22** $7 \times 4 = 28$ _____

23 $4 \times 5 = 20$ _____ **24** $2 \times 8 = 16$ _____

25 $6 \times 4 = 24$ _____ **26** $3 \times 9 = 27$ _____

27 $2 \times 11 = 22$ _____ **28** $10 \times 5 = 50$ _____

Solve.

29 **PACKAGING**   Three packages contain 36 golf balls in all. How many golf balls are in each package if there are the same number of balls in each?

_____

30 **MOVIES**   Eleanor received 18 movie passes for her birthday. How many times can she and 2 friends go to see a movie?

_____

Vocabulary Check   **Write the vocabulary word that completes each sentence.**

31 _____ is the same as repeated subtraction.

32 The _____ for division is multiplication.

33 **Writing in Math**   Karen wrote the division facts from the fact family $4 \times 5 = 20$. What mistake did Karen make?

$20 \div 5 = 5$ $\qquad\qquad$ $20 \div 4 = 4$

_____

_____

STOP

Divide by 0, 1, and 10

KEY Concept

There are special division rules to use when you divide.

Any number divided by itself is equal to 1.

$$3 \div 3 = 1$$

Any number divided by 1 is the same number.

$$3 \div 1 = 3$$

Zero divided by any number (except 0) equals 0.

You cannot divide by 0. It is not possible.

You can use models to divide by ten.

$$30 \div 10 = 3$$

Think: How many tens equal 30?

VOCABULARY

dividend
a number that is being divided

divisor
the number by which the dividend is being divided

quotient
the answer or result of a division problem

You should memorize the division rules for 0, 1, and 10.

- 0: Zero divided by any number equals zero.
- 1: Any number divided by one is the same number.
- 10: Models can be used to divide by ten.

GO ON

Example 1

Use a model to find 6 ÷ 1.

1. Draw a model.

2. How many groups will there be?   1

3. How many in each group?   6

4. Write the quotient.   6 ÷ 1 = 6

5. Check.   1 × 6 = 6

YOUR TURN!

Use a model to find 7 ÷ 1.

1. Draw a model.

2. How many groups will there be?   _____

3. How many in each group?   _____

4. Write the quotient.   7 ÷ 1 = _____

5. Check.   1 × 7 = _____

Example 2

Find 40 ÷ 10.

1. What number is the divisor? 10

2. How many tens equal 40?   4

3. Write the quotient.   40 ÷ 10 = 4

4. Check.   4 × 10 = 40

YOUR TURN!

Find 80 ÷ 10.

1. What number is the divisor?_____

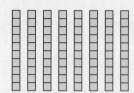

2. How many tens equal 80?   _____

3. Write the quotient. 80 ÷ 10 = _____

4. Check.   8 × 10 = _____

Who is Correct?

Find 100 ÷ 10.

Conchita
100 ÷ 10
= 10

John
100 ÷ 10
= 100

Lora
100 ÷ 10
= 1,000

Circle correct answer(s). Cross out incorrect answer(s).

▶ Guided Practice

Use a model to find each quotient.

1 5 ÷ 1 = _____

2 60 ÷ 10 = _____

Step by Step Practice

3 Find 50 ÷ 10.

Step 1 What number is the divisor? _____

Step 2 How many tens equal 50? _____

Step 3 Write the quotient. 50 ÷ 10 = _____

Step 4 Check. _____ × 10 = _____

GO ON

Find each quotient. If the quotient is not possible, write *not possible*.

4 90 ÷ 10 = _____

Check. _____ × _____ = _____

5 10 ÷ 10 = _____   **6** 70 ÷ 10 = _____

7 60 ÷ 10 = _____   **8** 40 ÷ 10 = _____

9 6 ÷ 1 = _____   **10** 9 ÷ 1 = _____

11 15 ÷ 0 = _____   **12** 4 ÷ 0 = _____

13 12 ÷ 1 = _____   **14** 3 ÷ 1 = _____

15 0 ÷ 7 = _____   **16** 0 ÷ 2 = _____

17 11 ÷ 11 = _____   **18** 8 ÷ 8 = _____

Step by Step **Problem-Solving Practice**

Solve.

19 **MONEY** There are 10 pennies in each dime. If you have 60 pennies, for how many dimes could you trade?

Understand	Read the problem. Write what you know.
	There are _____ pennies in a dime.
	There are _____ pennies altogether.
Plan	Pick a strategy. One strategy is to make a table.
Solve	Complete the table.
	The table shows that 60 pennies equal _____ dimes.
Check	Does the answer make sense? Look over your solution. Did you answer the question?

Problem-Solving Strategies

☐ Draw a model.
☐ Use logical reasoning.
☑ Make a table.
☐ Solve a simpler problem.
☐ Work backward.

Pennies	Dimes
10	1
20	2

20 **CONSTRUCTION** There are 10 rolls of paper towels in every jumbo package. If a shopper wants 80 rolls of paper towels, how many packages should he purchase? Check off each step.

_____ Understand: I circled key words.

_____ Plan: To solve the problem, I will _____.

_____ Solve: The answer is _____.

_____ Check: I checked my answer by _____.

21 **FOOD** The cafeteria used 120 eggs for a recipe. How many dozens of eggs did they use? Hint: use base-ten blocks.

_____

A dozen is **12** eggs.

22 **Reflect** Explain why you cannot divide by zero.

_____

_____

_____

▶ Skills, Concepts, and Problem Solving

Use a model to find each quotient.

23 $8 \div 8 =$ _____

24 $50 \div 10 =$ _____

25 $60 \div 10 =$ _____

26 $40 \div 10 =$ _____

GO ON

Find each quotient. If the quotient is not possible, write *not possible*.

27 $60 \div 10 =$ _____

28 $90 \div 1 =$ _____

29 $12 \div 1 =$ _____

30 $15 \div 1 =$ _____

31 $40 \div 0 =$ _____

32 $40 \div 1 =$ _____

33 $30 \div 10 =$ _____

34 $0 \div 80 =$ _____

35 $3 \div 0 =$ _____

36 $0 \div 7 =$ _____

37 $1 \div 1 =$ _____

38 $2 \div 1 =$ _____

39 $8 \div 1 =$ _____

40 $4 \div 1 =$ _____

41 $3 \div 3 =$ _____

42 $5 \div 5 =$ _____

Solve.

43 **MODELS**   Alana uses base-ten blocks to model a division problem. She takes the one-blocks and groups them into 10 piles of 9 blocks. What division sentence models her actions?

_____

44 **INTERIOR DESIGN**   Larry is laying tiles on the kitchen floor. The tiles come in boxes of 10. If each tile is 1 square foot, how many boxes will Larry need to tile a floor that is 80 square feet?

_____

45 **STORAGE**   Each storage rack in the school gym can hold 12 basketballs. Mrs. Booker has 12 basketballs. How many storage racks does Mrs. Booker need? What division sentence models her actions?

_____

46 **FITNESS**   Jake's grandfather started training on an exercise machine. By the third month, he trained 10 times longer each day than he did the first month. He trained for 60 minutes a day the third month. How many minutes a day did he train when he began?

_____

Vocabulary Check   **Write the vocabulary word that completes each sentence.**

47 In a division problem, the number being divided is the

_____ .

48 Zero cannot be a _____ because you cannot divide by zero.

49 **Writing in Math**   Explain why division can be thought of as repeated subtraction.

_____

_____

_____

▶ Spiral Review

Solve. (Lesson 3-1, p. 92)

50 **HOBBIES**   Ti is planting 35 flowers. He wants to have 5 rows of flowers. How many flowers will he put in each row?

_____

51 **BOOKS**   Stella read a total of 6 books over the 12 weeks of summer. Terrance read a total of 20 books during the 40 weeks of the school year. Who read faster? Explain your answer.

_____

_____

Progress Check 1 (Lessons 3-1 and 3-2)

Draw a model for each expression.

1 8 ÷ 2

2 15 ÷ 5

Write the division expression represented by each model.

3 _____

4 _____

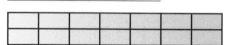

Find each quotient. If the quotient is not possible, write *not possible*.

5 60 ÷ 10 = _____

6 70 ÷ 10 = _____

7 10 ÷ 1 = _____

8 30 ÷ 1 = _____

9 20 ÷ 0 = _____

10 70 ÷ 0 = _____

11 50 ÷ 50 = _____

12 35 ÷ 35 = _____

13 0 ÷ 6 = _____

14 0 ÷ 10 = _____

Solve.

15 **MUSIC** Tina plays a series of 7 notes over and over. If she plays a total of 70 notes, how many times does Tina play the series? _____

16 **PARTIES** Jeannine placed 30 items into 3 favor bags. She placed the same number of items in each bag. How many items did she place in each bag? _____

17 **HISTORY** Refer to the photo caption at the right. A decade is 10 years. How many decades ago was the Declaration of Independence signed? _____

HISTORY The Declaration of Independence was signed about 230 years ago.

Divide by 2 and 5

KEY Concept

There are several phrases that represent dividing by 2.
For example, 8 ÷ 2 can mean:

- the quotient of 8 and 2
- 8 divided by 2
- half of 8

You can use money and related multiplication facts when dividing by 5.

There is 25¢ in nickels shown. Each nickel is worth 5¢.
There are 5 nickels. So, 25 ÷ 5 = 5. You can check using multiplication. 5 × 5 = 25.

Memorize the related multiplication facts of 2 and 5.

Example 1

Draw an array to find 8 ÷ 2.

1. Draw an array.

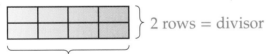

2 rows = divisor

4 columns = quotient

2. How many rectangles will be in the array?   **8**

3. How many rows?   **2**

4. Count the number of columns.   **4**

5. Write the quotient.   **8 ÷ 2 = 4**

6. Check.   **4 × 2 = 8**

GO ON

YOUR TURN!

Draw an array to find 12 ÷ 2.

1. Draw an array.

2. How many rectangles will be in the array? _____

3. How many rows? _____

4. Count the number of columns. [array image] } _____ rows = divisor

_____

_____ columns = quotient

5. Write the quotient.

   12 ÷ 2 = _____

6. Check. _____ × 2 = 12

Example 2

Find 40 ÷ 5.

1. Write the answer if you know it. Otherwise, solve using related multiplication.

2. What number multiplied by 5 equals 40? **8**

3. Write the quotient. **40 ÷ 5 = 8**

4. Check. How many nickels equal 40¢?
   8 nickels

YOUR TURN!

Find 35 ÷ 5.

1. Write the answer if you know it. Otherwise, solve using related multiplication.

2. What number multiplied by 5 equals 35?

   _____

3. Write the quotient.

   _____ ÷ _____ = _____

4. Check. How many nickels equal 35¢?

   _____

Who is Correct?

Find 45 ÷ 5.

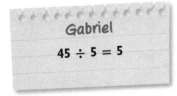

Gabriel

45 ÷ 5 = 5

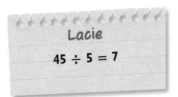

Lacie

45 ÷ 5 = 7

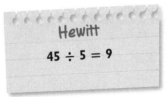

Hewitt

45 ÷ 5 = 9

Circle correct answer(s). Cross out incorrect answer(s).

▶ Guided Practice

Draw a model and find each quotient.

1 $15 \div 5 = $ _____

number of columns = _____ = quotient

2 $12 \div 2 = $ _____

number of columns = _____ = quotient

Use multiplication facts to help you.

Find each quotient.

3 $30 \div 5 = $ _____

4 $45 \div 5 = $ _____

5 $18 \div 2 = $ _____

6 $14 \div 2 = $ _____

7 $50 \div 5 = $ _____

8 $20 \div 2 = $ _____

9 $20 \div 5 = $ _____

10 $16 \div 2 = $ _____

11 $8 \div 2 = $ _____

Step by Step Practice

12 Find $60 \div 5$.

Step 1 Write the answer if you know it. Otherwise, solve using related multiplication.

Step 2 Write the related multiplication fact.

$5 \times $ _____ $= 60$

Step 3 Write the quotient.

_____ $\div$ _____ $=$ _____

Step 4 Check.

_____ $\times$ _____ $=$ _____

GO ON

Find each quotient.

13 35 ÷ 5 _____
Check your answer. _____ × _____ = _____

14 12 ÷ 2 _____

15 20 ÷ 2 _____

16 15 ÷ 5 _____

17 55 ÷ 5 _____

Step by Step Problem-Solving Practice

Solve.

18 **FASHION** If you have 28 socks, how many pairs of socks do you have?

Understand	Read the problem. Write what you know. There are _____ socks in each pair of socks. There are _____ socks.
Plan	Pick a strategy. One strategy is to draw a model.
Solve	Draw 28 tallies to represent the individual socks. Circle each pair. How many groups did you have?

Problem-Solving Strategies
☑ Draw a model.
☐ Use logical reasoning.
☐ Make a table.
☐ Solve a simpler problem.
☐ Work backward.

FASHION There are 2 socks in each pair.

There are _____ pairs of socks.

Check Use multiplication to check your answer.

19 **MONEY** There are $5 in every 5-dollar bill. Felix has $45 in 5-dollar bills. How many 5-dollar bills does Felix have? Check off each step.

_____ Understand: I circled key words.

_____ Plan: To solve the problem, I will _____.

_____ Solve: The answer is _____.

_____ Check: I checked my answer by _____.

20 **FITNESS**   Anica jogs 16 miles in 2 days. Aleesha jogs 40 miles in 5 days. Both jog an equal distance each day. Who jogs more each day?

_____

21   Reflect   Explain how using money can help you divide by 5.

_____

_____

▶ Skills, Concepts, and Problem Solving

Draw a model to find each quotient.

22 $35 \div 5$ _____

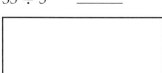

23 $44 \div 2$ _____

Find each quotient.

24 $18 \div 2 =$ _____

25 $12 \div 2 =$ _____

26 $15 \div 5 =$ _____

27 $25 \div 5 =$ _____

28 $45 \div 5 =$ _____

29 $20 \div 5 =$ _____

30 $14 \div 2 =$ _____

31 $18 \div 2 =$ _____

32 $25 \div 5 =$ _____

33 $50 \div 5 =$ _____

34 $22 \div 2 =$ _____

35 $14 \div 2 =$ _____

Solve.

36 **COMMUNITY SERVICE**   The sixth grade is collecting box tops to donate to a charity that will trade them in for cash. The students have 5 weeks to collect. If their goal is to collect 50 box tops, how many should they collect per week?

_____

GO ON

37 **FOOD**   Mrs. Foster made 2 pans of meatloaf and 5 pans of macaroni and cheese to serve for lunch. The meatloaf is cut so that there are 20 servings altogether. The macaroni and cheese is cut so there are 30 servings total. Which item has more servings in 1 pan?

_____

38 **SPORTS**   The after-school sports club rented the gym. It cost $60. If 5 sponsors donated an equal amount of money to cover the rental fees, then how much did each sponsor contribute?

_____

Vocabulary Check   **Write the vocabulary word that completes each sentence.**

39 If you divide something in _____, it is divided into two equal parts.

40 In a division problem, the number you divide by is the _____.

41 **Writing in Math**   What mistake did Mario make in dividing 17 by 2 and getting 9?

_____

 Spiral Review

Solve. (Lesson 3-2, p. 99)

42 **GRAPHING**   While making a scale for her graph, Sarah wants each side of each square to represent 10 years. If she plots a point that represents 70 years, how many squares high will it be?

_____

43 **FOOD**   The cafeteria manager wants to ensure there is 1 piece of pizza for every student. If he cuts each pizza into 10 slices, how many whole pizzas will he need to have enough for 120 students?

_____

44 **PACKAGING**   Cinnamon rolls come in packages of 7 or 8. What combination of packages is needed for exactly 30 cinnamon rolls?

_____

STOP

Divide by 3 and 4

KEY Concept

When a number does not divide evenly, the part left over is called the **remainder**. Suppose 3 friends share 7 cookies.

There is 1 cookie left over. This is the remainder.

When there are remainders, **multiples** help you find the number closest to the dividend.

The cookie example shows $7 \div 3$. The closest multiple of 3 is 6, leaving 1 left over.

VOCABULARY

multiple
a multiple of a number is the *product* of that number and any whole number
Example: 30 is a multiple of 10 because $3 \times 10 = 30$.

remainder
the number that is left after one whole number is divided by another

Practice memorizing the division facts for 3 and 4.

Example 1

Draw an array to model $8 \div 4$.

1. Write the answer if you know it. Otherwise, draw an array.

2. How many rectangles will be in the array? **8**

3. How many rows? **4**

2 columns = quotient

4. Count the number of columns. **2**

5. Write the quotient. $8 \div 4 = 2$

6. Check. $2 \times 4 = 8$

YOUR TURN!

Draw an array to model $20 \div 4$.

1. Write the answer if you know it. Otherwise, draw an array.

2. How many rectangles will be in the array? _____

3. How many rows? _____

4. Count the number of columns. _____

5. Write the quotient. $20 \div 4 =$ _____

6. Check. _____ $\times 4 = 20$

GO ON

Example 2

Find 8 ÷ 3. Show the remainder.

1. Rewrite the problem in vertical format.

$$3\overline{)8}$$

2. What number multiplied by 3 is close to 8?

3×3 is 9. That is too much.
3×2 is 6. This is close without going over.

$$\begin{array}{r} 2 \\ 3\overline{)8} \end{array}$$

Multiply. Write the product under the dividend.

$$\begin{array}{r} 2 \\ 3\overline{)8} \\ -6 \\ \hline 2 \end{array}$$

3. Subtract.

4. There are no more digits. The quotient is 2 with a remainder of 2. Write 2 R2.

$$\begin{array}{r} 2\ R2 \\ 3\overline{)8} \\ -6 \\ \hline 2 \end{array}$$

5. Check your answer. Multiply the quotient by the divisor. Then add the remainder.

$2 \times 3 = 6$
$6 + 2 = 8$

YOUR TURN!

Find 13 ÷ 4. Show the remainder.

1. Rewrite the problem in vertical format.

$$\overline{)}$$

2. What number multiplied by 4 is close to 13?

$4 \times \underline{\hspace{1cm}} = \underline{\hspace{1cm}}$

$$4\overline{)13}$$

Multiply. Write the product under the dividend.

$$4\overline{)13}$$

3. Subtract.

4. The quotient is _____ with a remainder of _____.

$$4\overline{)13}$$

5. Check your answer.

_____

_____

Who is Correct?

Find 22 ÷ 3. Show the remainder.

Zoe
$3 \times 7 = 21$
$21 + 1 = 22$
7 R1

Howie
$$\begin{array}{r} 6\ R4 \\ 3\overline{)22} \\ -18 \\ \hline 4 \end{array}$$

Tyrone
$$\begin{array}{r} 7\ R1 \\ 3\overline{)22} \\ -21 \\ \hline 1 \end{array}$$

Circle correct answer(s). Cross out incorrect answer(s).

▶ Guided Practice

Draw an array to model each quotient.

1 18 ÷ 3

number of columns = _____ = quotient

2 12 ÷ 4

number of columns = _____ = quotient

3 20 ÷ 4

number of columns = _____ = quotient

4 15 ÷ 3

number of columns = _____ = quotient

Step by Step Practice

5 Find 13 ÷ 3. Show the remainder.

Step 1 Rewrite the problem in vertical format.

Look at the first digit in the dividend. Since the divisor is _____ than the first digit, look at the first two digits.

$$\overline{)}$$

Step 2 What number multiplied by 3 is equal to or less than 13? _____

Multiply _____ × _____.

Write the answer under the dividend.

$$\overline{)}$$

Step 3 Subtract.

Step 4 There are no more digits. Write the quotient.

13 ÷ 3 = _____

Step 5 Check your answer.

_____ × _____ = _____

_____ + _____ = _____

GO ON →

Find each quotient. Show the remainder.

6 $14 \div 4$ _____

$4\overline{)14}$

Check your answer. _____ × _____ = _____

_____ + _____ = _____

7 $11 \div 3 =$ _____

8 $23 \div 3 =$ _____

9 $18 \div 4 =$ _____

10 $21 \div 4 =$ _____

Problem-Solving Strategies
☐ Draw a diagram.
☑ Use logical reasoning.
☐ Make a table.
☐ Solve a simpler problem.
☐ Work backward.

Step (by) Step **Problem-Solving Practice**

Solve.

11 **TRIPS** Twenty-one students are going on a field trip in three vans. One-third of the students will ride in each van. How many students will ride in each van?

One-third means to divide by 3.

Understand Read the problem. Write what you know.

There are _____ students.

_____ of _____ will ride in each van.

Plan Pick a strategy. One strategy is to use logical reasoning.

One-third means that a whole is divided into _____ equal parts.

Solve Divide 21 by 3.

_____ students will ride in each van.

Check Check your answer with repeated addition.

_____ + _____ + _____ = _____

TRIPS Twenty-one students will go on the field trip.

12 EGGS A farmer collected 8 eggs from 4 of his chickens. If each chicken produced the same number of eggs, how many eggs did each chicken produce? Check off each step.

_____ **Understand: I circled key words.**

_____ **Plan: To solve the problem, I will** _____.

_____ **Solve: The answer is** _____.

_____ **Check: I checked my answer by** _____.

13 MONEY If Andre has 20 quarters, how many dollars does he have? (There are 4 quarters in a dollar.)

_____

14 Reflect There are different ways to interpret a remainder. Suppose there are 25 students who are to sit in chairs that will be arranged in rows with 4 seats.

$25 \div 4 =$ _____.

How would you interpret the remainder?

_____

_____

_____

▶ Skills, Concepts, and Problem Solving

Draw an array to model each quotient.

15 $44 \div 4$

16 $24 \div 3$

17 $15 \div 3$

18 $16 \div 4$

GO ON

Find each quotient. Show the remainder if there is one.

19 $16 \div 4 =$ _____   **20** $9 \div 3 =$ _____

21 $12 \div 3 =$ _____   **22** $20 \div 4 =$ _____

23 $27 \div 3 =$ _____   **24** $36 \div 4 =$ _____

25 $6 \div 3 =$ _____   **26** $32 \div 4 =$ _____

27 $15 \div 3 =$ _____   **28** $40 \div 4 =$ _____

29 $12 \div 4 =$ _____   **30** $8 \div 4 =$ _____

31 $8 \div 3 =$ _____   **32** $10 \div 3 =$ _____

33 $9 \div 4 =$ _____   **34** $20 \div 3 =$ _____

35 $35 \div 4 =$ _____   **36** $28 \div 3 =$ _____

Solve.

37 PETS   Four friends visit a pet store. They each buy the same number of fish. If they bought 28 fish altogether, how many did each friend buy?

_____

38 BUSINESS   There are 10 packs of paper in a case. Three teachers share a case evenly. How much paper will each teacher get?

_____

39 FITNESS   In 10 days, Alan walked a total of 30 miles. In the same number of days, Jennifer walked 40 miles. How much more did Jennifer walk per day than Alan?

_____

Vocabulary Check   **Write the vocabulary word that completes each sentence.**

40 A _____ of a number is the product of that number and any whole number.

41 The _____ is the number left after one whole number is divided by another.

42 **Writing in Math**   Explain how to check a division problem when it has a remainder. Use the division problem $15 \div 4$ to show the work.

_____

_____

_____

_____

 **Spiral Review**

Find each quotient.

Solve. (Lesson 3-3, p. 107)

43 $30 \div 5 =$ _____

44 $18 \div 2 =$ _____

45 $20 \div 2 =$ _____

46 $60 \div 5 =$ _____

47 **BOOKS**   Together, Karina and her friend read 18 books over the summer. If they each read the same number of books, how many books did they read?

_____

48 **FOOD**   Sheila's manager asked her to take a survey of the customers in the pizza parlor. She was to ask customers at every other table to name their favorite pizza. If there were 16 tables with customers in the restaurant, how many tables did she go to?

_____

 STOP

Draw a model to find each quotient.

1 14 ÷ 2 _____

2 18 ÷ 3 _____

Write the division expression represented by each model.

3 _____

4 _____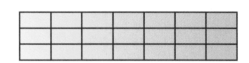

Find each quotient. Show the remainder if there is one.

5 35 ÷ 5 = _____ **6** 4 ÷ 2 = _____ **7** 4 ÷ 4 = _____ **8** 30 ÷ 3 = _____

9 50 ÷ 5 = _____ **10** 18 ÷ 3 = _____ **11** 36 ÷ 4 = _____ **12** 24 ÷ 3 = _____

13 30 ÷ 4 = _____ **14** 31 ÷ 5 = _____ **15** 11 ÷ 3 = _____ **16** 37 ÷ 4 = _____

Solve.

17 FARMS A dairy farmer has 40 cows. She began with 16 cows and bought the rest of her herd over the next 4 years. Not including her initial herd, how many cows did she buy each year?

_____

18 BIRDS The bird club counted 27 birds on an outing to the nature center. There are 3 acres in the nature center. How many birds are there per acre?

_____

19 PHOTOS Gordon's digital camera holds 43 pictures. Suppose he takes the same number of pictures each day. How many pictures can he take on a 4-day vacation? How many pictures will he have left?

_____

KEY Concept

You use the same process to divide by all single-digit numbers. Sometimes you will have remainders and sometimes you will not have remainders.

$$125 \text{ R4}$$
$$6)\overline{754}$$
$$-6$$
$$\overline{15}$$
$$-12$$
$$\overline{34}$$
$$-30$$
$$\overline{4}$$

$1 \times 6 = 6$

$2 \times 6 = 12$

$5 \times 6 = 30$

$$112$$
$$7)\overline{784}$$
$$-7$$
$$\overline{8}$$
$$-7$$
$$\overline{14}$$
$$-14$$
$$\overline{0}$$

$1 \times 7 = 7$

$1 \times 7 = 7$

$2 \times 7 = 14$

A remainder must always be less than the divisor.

VOCABULARY

multiple
a multiple of a number is the *product* of that number and any whole number
Example: 30 is a multiple of 10 because $3 \times 10 = 30$.

remainder
the number that is left after one whole number is divided by another

Memorize the division facts for 6 and 7.

Example 1

Write the division problem represented by the model.

1. How many rectangles are in the array? **18**

2. How many rows? **6**

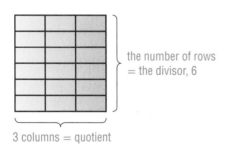

the number of rows = the divisor, 6

3 columns = quotient

3. Count the number of columns. This is the quotient. **3**

4. Write the division problem. **18 ÷ 6 = 3**

5. Check. **3 × 6 = 18**

YOUR TURN!

Write the division problem represented by the model.

1. How many rectangles are in the array? _____

2. How many rows? _____

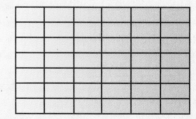

3. Count the number of columns. This is the quotient. _____

4. Write the division problem.
_____ ÷ _____ = _____

5. Check. _____ × _____ = _____

GO ON

Example 2

Find 696 ÷ 6.

1. Rewrite the problem in vertical format.

   $6\overline{)696}$

2. Look at the first digit. What number multiplied by 6 is 6? **1 × 6 = 6** Write 1 in the hundreds place in the quotient.

   $\overset{1}{6\overline{)696}}$

3. Multiply. Write the product under the hundreds place in the dividend.

4. Subtract. Bring down the next number in the dividend.

   $\begin{array}{r} 1 \\ 6\overline{)696} \\ -6\downarrow \\ \hline 9 \end{array}$

5. What number multiplied by 6 is close to 9, but not more than 9? **1 × 6 = 6** Write 1 in the tens place in the quotient.

   $\begin{array}{r} 11 \\ 6\overline{)696} \\ -6 \\ \hline 9 \\ -6 \\ \hline 3 \end{array}$

6. Multiply. Write the product under the dividend.

7. Subtract. Bring down the 6 in the ones place of the dividend.

8. What number multiplied by 6 is 36? **6** Multiply. Write the product under the dividend.

   $\begin{array}{r} 116 \\ 6\overline{)696} \\ -6 \\ \hline 9 \\ -6 \\ \hline 36 \\ -36 \\ \hline 0 \end{array}$

9. Subtract.

10. The quotient is 116.

YOUR TURN!

Find 784 ÷ 7.

1. Rewrite the problem in vertical format.

   $\overline{)}$

2. Look at the first digit. What number multiplied by 7 is 7? _____ Write _____ in the hundreds place of the quotient.

   $7\overline{)784}$

3. Multiply. Write the product under the hundreds place in the dividend.

4. Subtract. Bring down the next number in the dividend.

   $7\overline{)784}$

5. What number multiplied by 7 is close to 8, but not more than 8? _____ Write _____ in the tens place in the quotient.

   $7\overline{)784}$

6. Multiply. Write the product under the dividend.

7. Subtract. Bring down the next number in the ones place of the dividend.

8. What number multiplied by 7 is 14? _____. Multiply. Write the product under the dividend.

   $7\overline{)784}$

9. Subtract.

10. The quotient is _____.

Who is Correct?

Find 636 ÷ 7. Show the remainder.

Shelly
```
    9 R6
7)636
  −63
    6
```

Rasheeka
```
    90 R6
7)636
  −63
    6
```

Ellis
```
    91
7)636
  −63
    0
```

Circle correct answer(s). Cross out incorrect answer(s).

 Guided Practice

Write the division problem represented by the model.

1 _____

2 _____

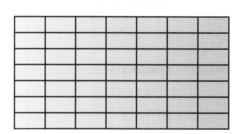

Step by Step Practice

3 Find 20 ÷ 6. Show the remainder.

Step 1 Rewrite the problem in vertical format.

```
6)20
```

Step 2 Look at the first digit in the dividend. Since the divisor is greater than the first digit, look at the first two digits. What number multiplied by 6 is close to 20? _____

Step 3 Multiply. Write the product under the dividend.

Step 4 Subtract.

Step 5 There are no more digits. Write the quotient.   20 ÷ 6 = _____

Step 6 Check.

_____

_____

GO ON

Find each quotient. Show the remainder if there is one.

4 15 ÷ 7 _____

$$7\overline{)15}$$

5 19 ÷ 6 = _____

6 35 ÷ 7 = _____

7 45 ÷ 7 = _____

8 39 ÷ 6 = _____

9 50 ÷ 7 = _____

10 54 ÷ 6 = _____

Step by Step **Problem-Solving Practice**

Solve.

11 HOBBIES   Han makes jewelry. He bought beads to make 7 necklaces. He will use an equal number of beads for each necklace. If he purchased 400 beads, how many beads can he use to make each necklace?

Understand	Read the problem. Write what you know. There are _____ beads to make _____ identical necklaces.
Plan	Pick a strategy. One strategy is to solve a simpler problem.
Solve	When you multiply 7 by 5, the product is _____. When you multiply 7 by 50, the product is _____. Subtract this from 400. _____ What number multiplied by 7 is close to 50 without going over? _____ Multiply this by 7 and subtract the product from 50. What is left is the _____. Add the factors you multiplied by 7. Along with the remainder, this is the answer. _____ + _____ = _____ R _____ So, Han can use _____ beads per necklace.
Check	Does the answer make sense? Look over your solution. Did you answer the question?

Problem-Solving Strategies

☐ Draw a diagram.
☐ Use logical reasoning.
☐ Make a table.
☑ Solve a simpler problem.
☐ Work backward.

12 **ART** The art museum has 532 pieces to place into 6 exhibit halls. How many pieces will go into each exhibit hall? Check off each step.

_____ **Understand: I circled key words.**

_____ **Plan: To solve the problem, I will** _____.

_____ **Solve: The answer is** _____.

_____ **Check: I checked my answer by** _____.

13 **HEALTH** There are 90 vitamins in a bottle. If Janelle takes 1 vitamin a day, how many weeks will a bottle of vitamins last?

_____

14 **Reflect** Complete the four sections below for $28 \div 7 = 4$.

Write the fact family.

Draw an array to model the division fact.

Write the fact in vertical and fraction forms.

Draw circles and tally marks to model the division fact.

 ## Skills, Concepts, and Problem Solving

Write the division problem represented by each model.

15 _____

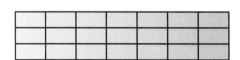

16 _____

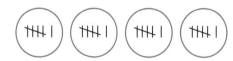

Write each quotient. Show the remainder if there is one.

17 $42 \div 6 =$ _____

18 $28 \div 7 =$ _____

19 $36 \div 6 =$ _____

20 $21 \div 7 =$ _____

21 $12 \div 6 =$ _____

22 $56 \div 7 =$ _____

GO ON

Write each quotient. Show the remainder if there is one.

23 $42 \div 7 =$ _____

24 $63 \div 7 =$ _____

25 $60 \div 6 =$ _____

26 $420 \div 7 =$ _____

27 $7 \div 7 =$ _____

28 $36 \div 6$ _____

29 $77 \div 7$ _____

30 $125 \div 6$ _____

31 $214 \div 7$ _____

Solve.

32 FOOD   There are blueberry and apple pies in the cafeteria. There are 42 pieces of each type of pie, but there is 1 more apple pie than blueberry pie. Each blueberry pie has 1 more piece than each apple pie. How many of each type of pie is there? How many pieces is each type of pie cut into?

_____

_____

Vocabulary Check   **Write the vocabulary word that completes each sentence.**

33 The _____ is the number that is left after one whole number is divided by another.

34 A(n) _____ of a number is the product of that number and any whole number.

35 Writing In Math   How does the inverse operation of division help you when solving division problems?

_____

_____

_____

 Spiral Review

Write each quotient.

36 $24 \div 3 =$ _____

37 $24 \div 4 =$ _____

38 $18 \div 3 =$ _____

39 $36 \div 4 =$ _____

STOP

Divide by 8 and 9

KEY Concept

As you become more comfortable with multiplication and division, try to use **mental math** as much as possible. Solving division problems using mostly mental math is called **short division**.

Ask yourself: 8 times what number equals 48? **8 × 6 = 48**

$$8\overline{)489} \quad \frac{61\ R1}{}$$

Ask yourself: 8 times what number equals 9 or a number very close to 9? **8 × 1 = 8**

As you practice short division, you will find ways to mark differences and remainders so that you do not have to do long division.

VOCABULARY

mental math
to add, subtract, multiply, and divide in your head without using manipulatives, fingers, or pencil and paper

multiple
a multiple of a number is the *product* of that number and any whole number
Example: 30 is a multiple of 10 because 3 × 10 = 30.

short division
division using mental math

Memorize the division facts for 8 and 9.

Example 1

Write the division problem represented by the model.

1. How many rectangles are in the array? **32**

2. How many rows? **8**

   The divisor equals the number of rows.

   4 columns = quotient

3. Count the number of columns. This is the quotient. **4**

4. Write the division problem.
   32 ÷ 8 = 4

5. Check.
   4 × 8 = 32

YOUR TURN!

Write the division problem represented by the model.

1. How many rectangles are in the array? _____

2. How many rows? _____

   The divisor equals the number of rows.

   _____ columns = quotient

3. Count the number of columns. This is the quotient. _____

4. Write the division problem.

   _____ ÷ _____ = _____

5. Check. _____ × _____ = _____   GO ON

Example 2

Find 289 ÷ 9. Show the remainder.

1. Rewrite the problem in vertical format.

$$9\overline{)289}$$

2. Look at the first digit. The divisor is greater than the first digit, so look at the first two digits. What number multiplied by 9 is close to 28? **3**

$$\begin{array}{r} 3 \\ 9\overline{)289} \\ -27 \\ \hline 1 \end{array}$$

Multiply. Write the product under the dividend. Now use mental math to subtract. **28 − 27 = 1**

3. Bring down the last digit. What is it? **9**

What number multiplied by 9 is close to 19? **2**

$$\begin{array}{r} 32 \\ 9\overline{)289} \\ -27\downarrow \\ \hline 19 \end{array}$$

4. Multiply. Write the product under the dividend. Then subtract.

$$\begin{array}{r} 32 \\ 9\overline{)289} \\ -27 \\ \hline 19 \end{array}$$

5. There are no more digits. The answer is 32 with a remainder of 1. Write 32 R1.

$$\begin{array}{r} 32 \\ 9\overline{)289} \\ -27 \\ \hline 19 \\ -18 \\ \hline 1 \end{array}$$

YOUR TURN!

Find 315 ÷ 8. Show the remainder.

1. Rewrite the problem in vertical format.

$$\overline{)}$$

2. Look at the first digit. The divisor is greater than the first digit, so look at the first two digits. What number multiplied by 8 is close to 31? _____

$$8\overline{)315}$$

Multiply. Write the product under the dividend. Now use mental math to subtract. 31 − 24 = _____

3. Bring down the last digit.

What number multiplied by 8 is close to 75? _____

$$8\overline{)315}$$

4. Multiply. Write the product under the dividend. Then subtract.

$$8\overline{)315}$$

5. There are no more digits. The answer is _____ with a remainder of _____.
Write _____ R _____.

Who is Correct?

Find 117 ÷ 9.

Danita

$$\begin{array}{r} 009 \\ 9\overline{)117} \\ -81 \\ \hline 36 \end{array}$$

Shannon

$$\begin{array}{r} 012 \\ 9\overline{)117} \\ -9 \\ \hline 27 \end{array}$$

Pedro

$$\begin{array}{r} 013 \\ 9\overline{)117} \\ -9 \\ \hline 27 \\ -27 \\ \hline 0 \end{array}$$

Circle correct answer(s). Cross out incorrect answer(s).

Guided Practice

Write the division problem represented by the model.

1 _____

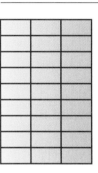

2 _____

Step (by) **Step Practice**

3 Find 128 ÷ 8. Use short division.

Step 1 Look at the first digit in the dividend. Because the 8 is greater than the 1, look at the first _____ digits in the dividend.

$$8\overline{)128}$$

Step 2 What number multiplied by 8 is close to 12 without going over? _____ Write the _____ in the tens place of the quotient.

Step 3 12 − _____ = _____. Write the difference next to the next digit in the dividend.

$$8\overline{)12\text{┆┄┆}8}$$

Step 4 What number will you divide now? _____

Step 5 What number multiplied by 8 is close to _____ without going over? _____ Write the _____ in the ones place of the quotient.

$$8\overline{)12\text{┆┄┆}8}$$

Step 6 The quotient is _____.

Check. _____ × 8 = 128

GO ON

Find each quotient. Show the remainder if there is one.

4 $108 \div 9 =$ $9\overline{)108}$ _____ × _____ = 108

5 $104 \div 8 =$ _____

6 $126 \div 9 =$ _____

7 $247 \div 8 =$ _____

8 $145 \div 9 =$ _____

Step by Step Problem-Solving Practice

Solve.

Problem-Solving Strategies
- ☐ Draw a diagram.
- ☐ Use logical reasoning.
- ☑ Make a table.
- ☐ Solve a simpler problem.
- ☐ Work backward.

9 **COMMUNITY SERVICE** The eleventh graders were repairing houses during spring break. For every 8 hours a student worked, he or she earned 1 class credit.

At the end of two weeks, the students worked a total of 720 hours. Each student worked an equal number of hours. If 10 students participated, how many credits did they earn in all?

Understand Read the problem. Write what you know.

There are _____ students. For every _____ hours worked, they received _____ credit. They worked _____ hours total.

Plan Pick a strategy. One strategy is to make a table.

Hours worked by each student	8	16	24							
Total hours for 10 students	80	160								
Credits earned	10									

Solve Look at the table. Find when the total hours worked is 720. The students earned _____ credits in all.

Check Does the answer make sense? Look over your solution. Did you answer the question?

Copyright © by The McGraw-Hill Companies, Inc.

10 CONSTRUCTION   A builder buys lumber by the linear foot. For each 9-foot length, she pays $5. How much does she pay for 81 linear feet of lumber? Check off each step.

_____ **Understand: I circled key words.**

_____ **Plan: To solve the problem, I will** _____.

_____ **Solve: The answer is** _____.

_____ **Check: I checked my answer by** _____

_____.

11 LANDSCAPING   The flower beds in front of school are placed in rows and columns of equal length. If there are 64 flowers planted in all, how many rows and how many columns are there?

_____

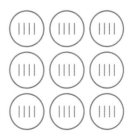

LANDSCAPING The flower beds are planted in rows and columns.

12 Reflect   Choose one fact family from the multiplication facts for 8 and another fact family for 9. Write all the facts for both families.

_____   _____

_____   _____

_____   _____

_____   _____

 Skills, Concepts, and Problem Solving

Write the division problem represented by the model.

13 _____

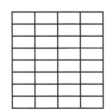

14 _____

(IIII) (IIII) (IIII)

(IIII) (IIII) (IIII)

Lesson 3-6 Divide by 8 and 9   **131**

Copyright © by The McGraw-Hill Companies, Inc.

Find each quotient.

15 $81 \div 9 = $ _____

16 $24 \div 8 = $ _____

17 $72 \div 9 = $ _____

18 $64 \div 8 = $ _____

19 $54 \div 9 = $ _____

20 $16 \div 8 = $ _____

21 $27 \div 9 = $ _____

22 $40 \div 8 = $ _____

23 $9 \div 9 = $ _____

24 $42 \div 7 = $ _____

25 $18 \div 9 = $ _____

26 $80 \div 8 = $ _____

27 $198 \div 9 = $ _____

28 $168 \div 8 = $ _____

29 $171 \div 9 = $ _____

30 $136 \div 8 = $ _____

Solve. Show your work.

SHOPPING   Bottled water costs $9 a case.

31 If Nalani has $80, how many cases of water can she buy?

_____

32 How many dollars will Nalani have left?

_____

SWIMMING   There are 85 students in the swim class. The game they are playing needs teams with 9 students on each team.

33 How many complete teams will there be?

_____

34 How many more students would be needed to make another team?

_____

SWIMMING There are 85 students in the swim class.

35 GAMES During 9 months, Da Jon paid $135 to play an online video game. If he paid the same amount each month, how much did he pay for 1 month?

_____

Vocabulary Check **Write the vocabulary word that completes each sentence.**

36 When you calculate a problem without using any tools (like paper and a pencil, or a calculator), you are performing

_____.

37 _____ is finding out how many times one number goes into another number using mental math.

38 Writing in Math Write two different ways that you could find the answer to 90 ÷ 10.

_____

_____

 Spiral Review

Find each quotient.

39 35 ÷ 7 = _____

40 18 ÷ 6 = _____

41 42 ÷ 6 = _____

42 63 ÷ 7 = _____

Solve. (Lesson 3-5, p. 121)

43 BUSINESS A store had sales of $1,750 in a 7-hour period. What were the sales per hour?

_____

44 FOOD Two cheese pizzas and two sausage pizzas were cut into 28 pieces. The cheese pizzas were cut into an equal number of pieces, and the sausage pizzas were cut into a different but equal number of pieces. How many pieces were in each pizza? Assume each pizza had over 5 pieces.

_____

Write the division problem represented by each model.

1 _____

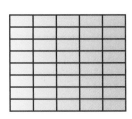

2 _____

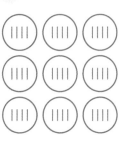

3 _____

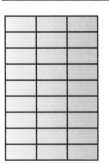

4 _____

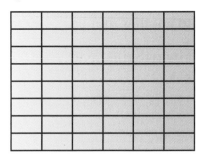

Find each quotient. Show a remainder if there is one.

5 $72 \div 8 =$ _____

6 $81 \div 9 =$ _____

7 $45 \div 9 =$ _____

8 $56 \div 8 =$ _____

9 $90 \div 8 =$ _____

10 $50 \div 9 =$ _____

11 $35 \div 9 =$ _____

12 $61 \div 8 =$ _____

13 $85 \div 8 =$ _____

14 $20 \div 8 =$ _____

15 $38 \div 9 =$ _____

16 $40 \div 9 =$ _____

Solve.

17 **SHOPPING** Paul is buying shirts. The shirts cost $8 each. If Paul has $43, how many shirts can he purchase? How much money will he have left?

_____

Vocabulary and Concept Check

divisor, *p. 92*

inverse operations, *p. 92*

multiple, *p. 113*

quotient, *p. 92*

remainder, *p. 113*

Write the vocabulary word that completes each sentence.

1 A(n) _____ of a number is the product of that number and any whole number.

2 _____ are operations that undo each other.

3 The number that follows the division sign in a division sentence is the _____.

4 A number that is left after one whole number is divided by another is a(n) _____.

Label each diagram below. Write the correct vocabulary term in each blank.

5 _____

$$\text{- - - - - →}\ \frac{24}{8} = 3$$

6 _____

$$\frac{77}{11} = 7$$
$$\text{- - - - - →}$$

7 _____

$$\text{- - - - - →}\ 7$$
$$8\overline{)56}$$

Lesson Review

3-1 Model Division (pp. 92–98)

Draw an array to model each expression.

8 12 ÷ 3

9 16 ÷ 4

Write in two different formats.

10 55 ÷ 5 _____

11 20 ÷ 2 _____

> ### Example 1
>
> **Draw an array to model the expression 9 ÷ 3.**
>
> 1. Identify the dividend. **9**
>
>   This is the *total number of rectangles* in the array.
>
> 2. Identify the divisor. **3**
>
>   This is the number of *rows*.
>
> 3. Draw an array with 9 rectangles in 3 rows.
>
>
>
>   3 columns = quotient
>
> 4. The number of columns in the array is the quotient. **3**
>
> 5. Check by multiplying the quotient by the divisor. **3 × 3 = 9**

3-2 Divide by 0, 1, and 10
(pp. 99–105)

Find each quotient.

12 $3 \div 1 =$ _____

13 $4 \div 0 =$ _____

14 $0 \div 8 =$ _____

15 $6 \div 6 =$ _____

16 $80 \div 10 =$ _____

17 $100 \div 10 =$ _____

Example 2

Find $50 \div 10$.

1. What number is the divisor?   10

2. What number times 10 equals 50? 5

3. Write the quotient. $50 \div 10 = 5$

4. Check. $5 \times 10 = 50$

3-3 Divide by 2 and 5
(pp. 107–112)

Find each quotient.

18 $20 \div 2 =$ _____

19 $12 \div 2 =$ _____

20 $15 \div 5 =$ _____

21 $55 \div 5 =$ _____

Example 3

Find $25 \div 5$.

1. Solve using related multiplication.

2. What number multiplied by 5 is 25?   5

3. Write the quotient. $25 \div 5 = 5$

4. Check. $5 \times 5 = 25$

3-4 Divide by 3 and 4
(pp. 113–119)

Find each quotient.
Show the remainder.

22 $18 \div 3$ _____

23 $26 \div 3$ _____

24 $35 \div 4$ _____

25 $36 \div 4$ _____

Example 4

Find $14 \div 4$. Show the remainder.

1. Rewrite the problem in vertical format.

2. What number multiplied by 4 is equal
to or less than 14? 3
$4 \times 3 = 12$
This is close without going over.
Multiply. Write the product under the dividend.

$$\begin{array}{r} 3 \\ 4\overline{)14} \\ -12 \\ \hline 2 \end{array}$$

3. Since there are no more digits to bring
down, the answer is 3 R2.

4. Check. $(3 \times 4) + 2 = 14$

3-5 Divide by 6 and 7 (pp. 121–126)

Find each quotient. Show any remainders.

26 $48 \div 6 =$ _____

27 $578 \div 7 =$ _____

28 $522 \div 6 =$ _____

29 $56 \div 7 =$ _____

Example 5

Find 288 ÷ 6.

1. Rewrite the pattern in vertical format.

2. Look at the first two digits. What number multiplied by 6 is close to 28? **4**

   4×6 is 24. Put the 4 in the tens place in the quotient. Bring down next digit.

$$
\begin{array}{r}
4 \\
6\overline{)288} \\
-24 \\
\hline
48
\end{array}
$$

3. What number multiplied by 6 is 48? **8** Complete the division.

4. The quotient is 48.

$$
\begin{array}{r}
48 \\
6\overline{)288} \\
-24 \\
\hline
48 \\
-48 \\
\hline
0
\end{array}
$$

3-6 Divide by 8 and 9 (pp. 127–133)

Write the division problem represented by the model.

30 _____

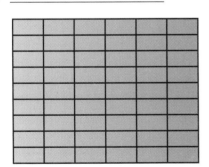

Find each quotient. Use short division.

31 $80 \div 8 =$ _____

32 $252 \div 9 =$ _____

33 $376 \div 8 =$ _____

34 $81 \div 9 =$ _____

Example 6

Write the division problem represented by the model.

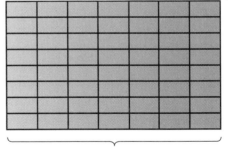

The divisor equals the number of rows.

7 columns = quotient

1. How many rectangles are in the array? **56**

2. How many rows? **8**

3. Count the number of columns. This is the quotient. **7**

4. Write the division problem. **56 ÷ 8 = 7**

5. Check. **8 × 7 = 56**

Write each expression in two different formats.

1   21 ÷ 7 _____

2   72 ÷ 8 _____

Write the division problem represented by the model.

3

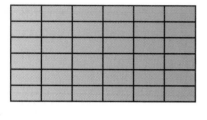

_____

4

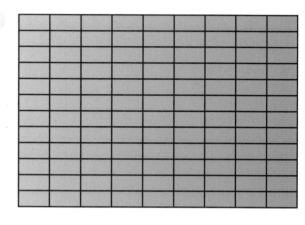

_____

5

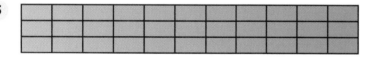

_____

Find each quotient. Show the remainder if there is one.

6   20 ÷ 10 = _____

7   11 ÷ 1 = _____

8   0 ÷ 5 = _____

9   60 ÷ 10 = _____

10   9 ÷ 3 = _____

11   8 ÷ 0 = _____

12   14 ÷ 7 = _____

13   63 ÷ 9 = _____

14   33 ÷ 3 = _____

15   0 ÷ 2 = _____

16   20 ÷ 10 = _____

17   35 ÷ 5 = _____

18   17 ÷ 5 = _____

19   51 ÷ 6 = _____

20   33 ÷ 4 = _____

GO ON ➡

Find each quotient. Show the remainder if there is one.

21 $58 \div 8 =$ _____

22 $57 \div 9 =$ _____

23 $36 \div 2 =$ _____

24 $45 \div 5 =$ _____

25 $64 \div 8 =$ _____

26 $324 \div 6 =$ _____

27 $24 \div 3 =$ _____

28 $72 \div 4 =$ _____

29 $49 \div 7 =$ _____

Solve.

SCHOOL   The principal wants to give a ribbon to every student who competes in the spelling bee. Ribbons are packaged 8 to a bundle.

30 If all 468 students compete, how many bundles of ribbons will the principal need?

_____

31 How many ribbons will be left over?

_____

Award ribbon

32 **LIBRARY**   The library has 672 books that need to be placed on 8 new shelves. If each shelf is to have the same number of books, how many books will be in each group?

_____

FOOD   The cafeteria needs a basket of 6 rolls per table for the banquet on Friday. They have baked 320 rolls.

33 How many baskets can be filled?

_____

34 How many extra rolls will there be after the baskets are filled?

_____

Correct the mistakes.

35 Orlando said that all fact families for multiplication and division have four equations. Give an example of a fact family with four equations. Give an example of a fact family with two equations that shows that Orlando's statement is false.

_____

_____

STOP

Choose the best answer and fill in the corresponding circle on the sheet at right.

1 Which equation can be used to check $10 \times 4 = 40$?

   A $40 \div 8 = 5$ **C** $4 \times 4 = 16$

   B $10 \times 10 = 100$ **D** $40 \div 10 = 4$

---

2 Chandani wants to share her music CD collection with her sister. If Chandani has 24 CDs and gives her sister half of her collection, how many CDs will each sister have?

   A 8 CDs   **C** 12 CDs

   B 10 CDs   **D** 20 CDs

---

3 Jeans are on sale this week. How much would one pair of jeans cost during this sale?

SALE
Jeans
3 for $54

   A $16   **C** $21

   B $18   **D** $162

---

4 The number that is left after one whole number is divided by another is the _____.

   A quotient   **C** remainder

   B dividend   **D** multiple

5 Ms. Wantobi is dividing counters equally into 9 jars for a math project. If she has a total of 765 counters, how many will go in each jar?

   A 73 counters

   B 84 counters

   C 85 counters

   D 91 counters

---

6 The music teacher wants to divide the band students into 8 equal groups. If there are 128 total students in band, which grouping will work?

   A 8 groups of 12 students

   B 9 groups of 15 students

   C 8 groups of 16 students

   D 11 groups of 10 students

---

7 The Lorenz family traveled 3,840 miles on their summer vacation. If they traveled the same number of miles per day and finished the trip in 8 days, how many miles per day did they travel?

   A 460 miles

   B 480 miles

   C 510 miles

   D 530 miles

GO ON

8 Karlie reads about 40 pages per hour. If she finishes her book in 9 hours, about how many pages are in this book?

    A 180 pages

    B 270 pages

    C 360 pages

    D 540 pages

---

9 Which of the following is not possible?

    A $12 \div 1$ **C** $12 \div 0$

    B $0 \div 12$ **D** $12 \div 12$

---

10 If Shaun has 40 quarters, how many dollars does he have?

    A $5

    B $10

    C $15

    D $20

---

11 Enrique is baking cookies. He wants to make 153 large cookies to sell for a school fundraiser. If he can make 9 cookies per batch, how many batches will he need to bake?

    A 14 batches

    B 15 batches

    C 16 batches

    D 17 batches

ANSWER SHEET

Directions: Fill in the circle of each correct answer.

1 Ⓐ Ⓑ Ⓒ Ⓓ
2 Ⓐ Ⓑ Ⓒ Ⓓ
3 Ⓐ Ⓑ Ⓒ Ⓓ
4 Ⓐ Ⓑ Ⓒ Ⓓ
5 Ⓐ Ⓑ Ⓒ Ⓓ
6 Ⓐ Ⓑ Ⓒ Ⓓ
7 Ⓐ Ⓑ Ⓒ Ⓓ
8 Ⓐ Ⓑ Ⓒ Ⓓ
9 Ⓐ Ⓑ Ⓒ Ⓓ
10 Ⓐ Ⓑ Ⓒ Ⓓ
11 Ⓐ Ⓑ Ⓒ Ⓓ

Success Strategy

Read the entire question before looking at the answer choices. Watch for words like *not* that change the whole question.

STOP

Chapter 4

Properties of Operations

Knowing the basics of math can help you do things like save for a trip. For example, how much would it cost your family to fly to Utah if each plane ticket costs $120 and there are 4 people in your family?

STEP 1 Quiz

Are you ready for Chapter 4? Take the Online Readiness Quiz at *glencoe.com* to find out.

STEP 2 Preview

Get ready for Chapter 4. Review these skills and compare them with what you'll learn in this chapter.

What You Know	What You Will Learn
You know how to multiply. **Examples:** $7 \times 4 = 28$ $4 \times 7 = 28$ **TRY IT!** 1 $5 \times 5 = $ _____ 2 $10 \times 10 = $ _____ 3 $7 \times 12 = $ _____ 4 $2 \times 24 = $ _____	*Lesson 4-1* The order in which you multiply numbers does not change the answer. This is called the **Commutative Property of Multiplication**. $7 \times 4 = 28$ $4 \times 7 = 28$ The answer is the same.
You know how to add. **Examples:** $4 + 3 + 2 = 9$ $2 + 3 + 4 = 9$ **TRY IT!** 5 $3 + 5 + 10 = $ _____ 6 $10 + 5 + 3 = $ _____ 7 $20 + 3 + 3 = $ _____ 8 $3 + 20 + 3 = $ _____	*Lesson 4-2* Grouping numbers you are adding in different ways does not change the answer. This is called the **Associative Property of Addition**. $(4 + 3) + 2 = 9$ $7 \quad + 2 = 9$ $4 + (3 + 2) = 9$ $4 + \quad 5 \quad = 9$
You know how to add and multiply. **Example:** $(2 \times 4) + (2 \times 5) = 18$ $8 \quad + \quad 10 \quad = 18$ **TRY IT!** 9 $(5 \times 2) + (3 \times 3) = $ _____ 10 $(10 \times 9) + (1 \times 4) = $ _____ 11 $(6 \times 3) + (4 \times 1) = $ _____ 12 $(6 \times 3) + (4 \times 2) = $ _____	*Lesson 4-3* Distribute a number by multiplying it with each member of a group. This is called the **Distributive Property**. $2(4 + 5) = $ $(2 \times 4) + (2 \times 5) = $ $8 \quad + \quad 10 \quad = 18$

Commutative Property

KEY Concept

The Commutative Property states that the order in which two numbers are added or multiplied does not change their sum or product.	
Property	**Example**
Commutative Property of Addition	$3 + 5 = 5 + 3$ $8 = 8$
Commutative Property of Multiplication	$4 \times 5 = 5 \times 4$ $20 = 20$

VOCABULARY

Commutative Property of Addition
   the order in which two numbers are added does not change the *sum*

Commutative Property of Multiplication
   the order in which two numbers are multiplied does not change the *product*

product
   the answer to a multiplication problem

sum
   the answer to an addition problem

The Commutative Property tells you that order does not matter when you are adding or multiplying.

Example 1

Draw a model to show $2 + 4 = 4 + 2$. Which property did you show?

1. Create a model for each side of the equation.

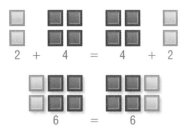

2 + 4 = 4 + 2

6 = 6

2. The order of the numbers changed, but the sum did not. This is the Commutative Property of Addition.

YOUR TURN!

Draw a model to show $3 \times 2 = 2 \times 3$. Which property did you show?

1. Create a model for each side of the equation.

2. Which property did you show?

_____

Example 2

Use the Commutative Property to fill in the blank. Check your answer.

$8 \times 6 =$ _____ $\times 8$

1. Use the Commutative Property of Multiplication.

$8 \times 6 = 6 \times 8$

2. Check by multiplying the numbers on each side of the equation.

$8 \times 6 = 6 \times 8$

$48 = 48$

YOUR TURN!

Use the Commutative Property to fill in the blank. Check your answer.

$9 + 3 =$ _____ $+ 9$

1. Use the Commutative Property of Addition.

$9 + 3 =$ _____ $+ 9$

2. Check by adding the numbers on each side of the equation.

_____

_____

Who is Correct?

Give an example of the Commutative Property of Multiplication.

Ira
$7 \times 2 = 2 \times 7$
$14 = 14$

Cynthia
$7 \times 2 = 14$

Diego
$7 \times 2 = 14 \times 1$
$14 = 14$

Circle correct answer(s). Cross out incorrect answer(s).

 ## Guided Practice

Draw a model to show each equation.

1. $2 + 5 = 5 + 2$
Which property did you show?

_____

2. $3 \times 4 = 4 \times 3$
Which property did you show?

_____

GO ON

Step by Step Practice

Use the Commutative Property to fill in the blank. Check your answer.

3 $12 + 9 = 9 + ____$

Step 1 The order in which two numbers are ____________ does not change the sum. The property shown is the ____________ Property of ____________.

Step 2 Which number is not shown on the right side of the equation? Fill in the blank.

$12 + 9 = 9 + ____$

Step 3 Check. Add the numbers on each side of the equation.

$12 + 9 = 9 + ____$

$21 = ____$

Use the Commutative Property to fill in each blank. Check your answer.

4 $3 \times 9 = ____ \times 3$

$27 = ____$

5 $____ + 5 = 5 + 18$

$____ = 23$

6 $8 + 11 = ____ + ____$

$____ = ____$

7 $5 \times 6 = ____ \times ____$

$____ = ____$

8 $4 + ____ = 7 + ____$

$____ = ____$

9 $2 \times ____ = 6 \times ____$

$____ = ____$

Draw a model to show each equation.

10 $2 \times 5 = 5 \times 2$

11 $7 + 3 = 3 + 7$

Solve.

12 **SHOPPING** Jacob bought 3 boxes of pens with 5 pens in each box. Lydia bought 5 boxes of pens with 3 pens in each box. Compare the number of pens Jacob and Lydia bought. Justify your answer.

Problem-Solving Strategies
☐ Draw a diagram.
☐ Look for a pattern.
☐ Guess and check.
☑ Act it out.
☐ Solve a simpler problem.

Understand   Read the problem. Write what you know.

Jacob bought _____ boxes with _____ pens each.

Lydia bought _____ boxes with _____ pens each.

Plan   Pick a strategy. One strategy is to act it out.

Arrange pens in rows and columns to show the pens that Jacob and Lydia bought. Then, write an expression to model each arrangement.

Solve   Jacob's arrangement is _____ rows by _____ columns.

He has _____ pens.

The expression is _____.

Lydia's arrangement is _____ rows by _____ columns.

She has _____ pens.

The expression is _____.

The number of pens that Jacob bought is _____ the number of pens that Lydia bought.

Check   Multiply the numbers on each side of the equation.

$3 \times 5 =$ _____ $\times$ _____

_____ $=$ _____

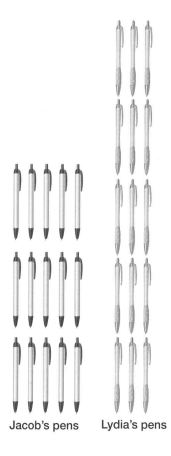

Jacob's pens   Lydia's pens

GO ON

13 **JEWELRY** Missy has 7 boxes of necklaces with 3 necklaces in each box. Sari has 3 boxes of necklaces with 7 necklaces in each box. Compare the number of necklaces. Justify your answer. Check off each step.

_____ **Understand: I circled key words.**

_____ **Plan: To solve the problem, I will** _____.

_____ **Solve: The answer is** _____.

_____ **Check: I checked my answer by** _____

_____.

14 **COIN COLLECTING** Billy has 35 coins. His brother gave him 62 more coins. Roberto has 62 coins. His father gave him 35 more coins. Compare the number of coins. Justify your answer.

_____

15 **STORES** The Corner Store had 15 sweaters on display. Ten more sweaters were delivered in the afternoon. The Sweater Store had 10 sweaters on display. In the afternoon delivery, 15 more sweaters arrived. Compare the number of sweaters. Justify your answer.

_____

16 **Reflect** How do you know that $65 + 98 = 98 + 65$ without adding?

_____

_____

_____

 ## Skills, Concepts, and Problem Solving

Draw a model to show each equation.

17 $6 \times 5 = 5 \times 6$
Which property did you show?

_____

18 $4 + 5 = 5 + 4$
Which property did you show?

_____

Use the Commutative Property to fill in each blank. Check your answer.

19 $7 + 6 = \underline{\hspace{1cm}} + 7$

$\underline{\hspace{1cm}} = 13$

20 $39 + 28 = 28 + \underline{\hspace{1cm}}$

$67 = \underline{\hspace{1cm}}$

21 $5 \times 9 = 9 \times \underline{\hspace{1cm}}$

$45 = \underline{\hspace{1cm}}$

22 $18 \times 5 = \underline{\hspace{1cm}} \times 18$

$\underline{\hspace{1cm}} = 90$

23 $48 + 37 = \underline{\hspace{1cm}} + \underline{\hspace{1cm}}$

$\underline{\hspace{1cm}} = \underline{\hspace{1cm}}$

24 $\underline{\hspace{1cm}} + \underline{\hspace{1cm}} = 15 + 11$

$\underline{\hspace{1cm}} = \underline{\hspace{1cm}}$

25 $\underline{\hspace{1cm}} \times \underline{\hspace{1cm}} = 7 \times 6$

$\underline{\hspace{1cm}} = \underline{\hspace{1cm}}$

26 $9 \times 8 = \underline{\hspace{1cm}} \times \underline{\hspace{1cm}}$

$\underline{\hspace{1cm}} = \underline{\hspace{1cm}}$

27 $5 + \underline{\hspace{1cm}} = 26 + \underline{\hspace{1cm}}$

$31 = \underline{\hspace{1cm}}$

28 $\underline{\hspace{1cm}} + 44 = \underline{\hspace{1cm}} + 55$

$\underline{\hspace{1cm}} = 99$

29 $8 \times \underline{\hspace{1cm}} = \underline{\hspace{1cm}} \times 8$

$32 = \underline{\hspace{1cm}}$

30 $5 \times \underline{\hspace{1cm}} = 9 \times \underline{\hspace{1cm}}$

$45 = \underline{\hspace{1cm}}$

31 $2 \times \underline{\hspace{1cm}} = \underline{\hspace{1cm}} \times 2$

$\underline{\hspace{1cm}} = 54$

32 $\underline{\hspace{1cm}} \times 18 = \underline{\hspace{1cm}} \times 3$

$\underline{\hspace{1cm}} = 54$

Solve.

33 HOBBIES   Sarah has 5 bags of marbles with 10 marbles in each bag. Noah has 10 bags of marbles with 5 marbles in each bag. Compare the number of marbles. Justify your answer.

_____

34 SHAPES   Elian has 27 triangles and 15 squares. Mykia has 15 triangles and 27 squares. Compare the number of shapes. Justify your answer.

_____

GO ON

35 **GAMES** Felipe had 14 game cards. He bought 12 more game cards. Wanda had 12 game cards. She bought 14. Compare the number of game cards. Justify your answer.

_____

36 **MUSIC** Nicolas saved 23 songs to one jump drive and 35 to another. Eliza saved 35 songs to one jump drive and 23 to another. Compare the number of songs they saved. Justify your answer.

_____

37 **NUMBERS** If you know that 195 + 126 = 321,

what is the sum of 126 + 195? _____

38 **NUMBERS** If you know that 26 × 11 = 286,

what is the product of 11 × 26? _____

Vocabulary Check **Write the vocabulary word that completes each sentence.**

39 The Commutative Property of _____ states that the order in which two numbers are added does not change the sum.

40 The _____ is the answer to a multiplication problem.

41 The Commutative Property of _____ states that the order in which two numbers are multiplied does not change the product.

42 The answer to an addition problem is the _____.

43 **Writing in Math** Use the Commutative Property of Multiplication to rewrite the expression 8 × 7. Does this change the product? Explain.

_____

_____

_____

STOP

Associative Property

KEY Concept

The Associative Property states that the way in which three numbers are grouped when they are added or multiplied does not change their sum or product.

Property	Example
Associative Property of Addition	$(4 + 6) + 9 = 4 + (6 + 9)$ $10 + 9 = 4 + 15$ $19 = 19$
Associative Property of Multiplication	$(2 \times 3) \times 5 = 2 \times (3 \times 5)$ $6 \times 5 = 2 \times 15$ $30 = 30$

VOCABULARY

addend
any numbers or quantities being added together

Associative Property of Addition
the grouping of the *addends* does not change the *sum*

Associative Property of Multiplication
the grouping of the *factors* does not change the *product*

factor
a number that divides a whole number evenly

The **Associative Property** uses grouping with parentheses to make addition and multiplication easier.

Example 1

Draw a model to show $(2 + 3) + 4 = 2 + (3 + 4)$. Which property did you show?

1. Create a model for each side of the equation.

$(2 + 3) + 4 = 2 + (3 + 4)$

$5 + 4 = 2 + 7$

$9 = 9$

2. The grouping of the addends did not change the sum. This is the Associative Property of Addition.

YOUR TURN!

Draw a model to show $(1 \times 3) \times 2 = 1 \times (3 \times 2)$. Which property did you show?

1. Create a model for each side of the equation.

2. Which property did you show?

_____

GO ON

Example 2

Use the Associative Property to fill in the blank. Check your answer.

$(7 \times 5) \times 2 = 7 \times (\underline{\hspace{1cm}} \times 2)$

1. Use the Associative Property of Multiplication.
   $(7 \times 5) \times 2 = 7 \times (5 \times 2)$

2. Check by multiplying the numbers on each side of the equation.
   $(7 \times 5) \times 2 = 7 \times (5 \times 2)$
   $35 \times 2 = 7 \times 10$
   $70 = 70$

YOUR TURN!

Use the Associative Property to fill in the blank. Check your answer.

$(2 + 4) + 1 = 2 + (\underline{\hspace{1cm}} + 1)$

1. Use the Associative Property of Addition.
   $(2 + 4) + 1 = 2 + (\underline{\hspace{1cm}} + 1)$

2. Check by adding the numbers on each side of the equation.

   _____

   _____

   _____

Who is Correct?

Use the Associative and Commutative Properties to find the sum of **25 + 94 + 75** mentally.

> Add the numbers first that give you a 0 in the ones place. This will make it easier to mentally add the third number.

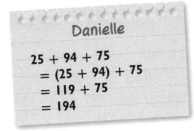

Danielle
$25 + 94 + 75$
$= (25 + 94) + 75$
$= 119 + 75$
$= 194$

Tamera
$25 + 94 + 75$
$= 25 + 75 + 94$
$= (25 + 75) + 94$
$= 100 + 94 \text{ or } 194$

Aaron
$25 + 94 + 75$
$= 25 + (94 + 75)$
$= 25 + 169$
$= 194$

Circle correct answer(s). Cross out incorrect answer(s).

 ## Guided Practice

Draw a model to show each equation.

1 $(1 + 2) + 3 = 1 + (2 + 3)$
   Which property did you show?

   _____

2 $(2 \times 3) \times 4 = 2 \times (3 \times 4)$
Which property did
you show?

_____

_____

Step by Step Practice

Use the Associative Property to find the sum mentally.

3 $8 + 4 + 16$

Step 1 Look for two numbers whose sum is 10, 20, 30, or another multiple of 10. The sum of _____ and _____ is a multiple of 10: _____.

Step 2 Rewrite the expression using the Associative Property.
$8 + 4 + 16 = 8 + (\_\_\_\_ + \_\_\_\_)$

> Add the numbers first that make it easier to mentally add the third number.

Step 3 Find the sum.
$8 + (4 + 16) = 8 + \_\_\_\_$ or $\_\_\_\_$

Use the Associative Property to fill in each blank. Check your answer.

4 $(9 \times 5) \times 2 = \_\_\_\_ \times (5 \times 2)$
$45 \times 2 = 9 \times \_\_\_\_$
$90 = \_\_\_\_$

5 $7 + (3 + 8) = (\_\_\_\_ + 3) + 8$
$7 + 11 = \_\_\_\_ + 8$
$18 = \_\_\_\_$

6 $5 \times 4 \times 3 = (\_\_\_\_ \times \_\_\_\_) \times \_\_\_\_$
$= \_\_\_\_ \times \_\_\_\_$
$= \_\_\_\_$

7 $(14 + 19) + 1 = \_\_\_\_ + (\_\_\_\_ + \_\_\_\_)$
$33 + 1 = \_\_\_\_ + \_\_\_\_$
$\_\_\_\_ = \_\_\_\_$

Use the Commutative and Associative Properties to find each sum or product mentally.

8 $9 + 21 + 30 = (\_\_\_\_ + \_\_\_\_) + \_\_\_\_$
$= \_\_\_\_ + \_\_\_\_$
$= \_\_\_\_$

9 $5 \times 21 \times 6 = \_\_\_\_ \times (\_\_\_\_ \times \_\_\_\_)$
$= \_\_\_\_ \times \_\_\_\_$
$= \_\_\_\_$

GO ON

Solve.

10 SCHOOL DAYS   Mr. Daniels sold 75 tickets to the school play on Monday, 52 tickets on Tuesday, and 48 tickets on Wednesday. How many tickets did Mr. Daniels sell in all? Explain your reasoning.

Understand   Read the problem. Write what you know.
The number of tickets sold was _____, _____, and _____.

In this problem, the words "in all" mean to _____.

Plan   Pick a strategy. One strategy is to solve a simpler problem.

Look for two numbers that will have a sum with a 0 in the ones place. Add those numbers together first.

Solve   Write an expression for how many tickets Mr. Daniels sold in all.

_____ + _____ + _____

Use the Associative Property of Addition to rewrite the expression so that it is easier to simplify mentally. Then find the sum.

_____ + _____ + _____ = (_____ + _____) + _____

= _____ + _____ or _____

Check   Use a calculator to find the sum.

11 STAMP COLLECTING   Uma has 44 stamps in her collection. Elena gave her 22 more stamps. Uma then bought 6 more stamps. How many stamps does Uma have in all? Explain your reasoning. Check off each step.

_____ Understand: I circled key words.

_____ Plan: To solve the problem, I will _____.

_____ Solve: The answer is _____.

_____ Check: I checked my answer by _____.

12 SPORTS Marty bought the boxes of whistles shown. Each whistle cost $2. How much did Marty spend? Explain your reasoning.

_____

_____

13 COOKING A casserole recipe calls for 2 packages of cheese. Marisela needs to make 3 casseroles. Each package of cheese costs $1.50. Find the cost of the cheese for all 3 casseroles. Explain your reasoning.

_____

14 **Reflect** Give an example of the Associative Property of Addition. Check your answer.

_____

_____

▶ Skills, Concepts, and Problem Solving

Draw a model to show each equation.

15 $(3 + 4) + 1 = 3 + (4 + 1)$
Which property did you show?

_____

_____

16 $(1 \times 5) \times 3 = 1 \times (5 \times 3)$
Which property did you show?

_____

_____

GO ON

Use the Associative Property to fill in each blank. Check your answer.

17 36 + (14 + 19) = (_____ + 14) + 19

              36 + 33 = _____ + 19

                 _____ = _____

18 8 × (5 × 4) = (_____ × 5) × 4

            8 × 20 = _____ × 4

              _____ = _____

19 6 × (5 × 3) = _____

               = _____ × _____

               = _____

20 18 + (22 + 37) = _____

               = _____ + _____

               = _____

21 (36 + 14) + 16 = _____

               = _____ + _____

               = _____

22 (63 + 13) + 17 = _____

               = _____ + _____

               = _____

Use the Commutative and Associative Properties to find each sum or product mentally.

23 15 × 7 × 2 = (_____ × _____) × _____

               = _____ × _____

               = _____

24 12 × 3 × 5 = (_____ × _____) × _____

               = _____ × _____

               = _____

25 102 + 89 + 18 = (_____ + _____) + _____

               = _____ + _____

               = _____

26 15 + 77 + 85 = (_____ + _____) + _____

               = _____ + _____

               = _____

Solve. Justify your answer.

27 **SHOPPING** Lana bought the packages of highlighters shown. Each highlighter cost $2. How much did Lana spend?

_____

_____

28 **CONSTRUCTION** Tregg has 17 nails. Isabel gave him 13 more nails. Tregg then bought 18 more nails. How many nails does Tregg have in all?

_____

29 **FOOD** Diana bought 4 packs of gum. Each pack contained 12 sticks of gum. The next week, she bought 2 more packs of gum. How many sticks of gum did Diana have in all?

_____

30 **NUMBERS** If you know that $3 + (5 + 4) = 12$, then what is the sum of $(3 + 5) + 4$?

_____

31 **NUMBERS** If you know that $2 \times (6 \times 11) = 132$, what is the product of $(2 \times 6) \times 11$?

_____

Vocabulary Check **Write the vocabulary word that completes each sentence.**

32 The Associative Property of _____ states that the grouping of the factors does not change the product.

33 The Associative Property of _____ states that the grouping of the addends does not change the sum.

34 **Writing in Math** Explain how you can use the Commutative and Associative Properties to help you find sums and products mentally.

_____

_____

_____

 Spiral Review

Solve. (Lesson 4-1, p. 144)

35 **SAFETY** Manuel has 5 boxes of safety glasses with 12 pairs of glasses in each box. Stephany has 12 boxes of safety glasses with 5 pairs of glasses in each box. Compare the number of glasses. Explain your reasoning.

_____

Solve.

36 $9 \times 3 = $ _____ $\times 9$

37 $8 \times 7 = 7 \times $ _____

38 $215 + $ _____ $= 150 + 215$

39 $14 + 23 = $ _____ $+ 14$

STOP

Draw a model to show 1 + 5 = 5 + 1. Which property did you show?

1 _____

_____

Draw a model to show (1 × 5) × 3 = 1 × (5 × 3). Which property did you show?

2 _____

_____

Use the Commutative Property to fill in each blank with the correct value. Check your answer.

3   $21 + 36 = $ _____ $+$ _____

_____ $=$ _____

4   $9 \times 3 = $ _____ $\times$ _____

_____ $=$ _____

Use the Commutative and Associative Properties to find each sum or product mentally.

5   $3 + 16 + 7 = ($ _____ $+$ _____ $) +$ _____

$= $ _____ $+$ _____ or _____

6   $5 \times 6 \times 3 = ($ _____ $\times$ _____ $) \times$ _____

$= $ _____ $\times$ _____ or _____

Solve. Justify your answer.

7   **FITNESS**   Brian does 30 minutes of aerobics on Tuesdays, 45 minutes on Thursdays, and 20 minutes on Saturdays. How many minutes of aerobics does Brian do altogether?

_____

8   **SHOPPING**   Lela bought the packages of pens shown. Each pen cost $2. How much did Lela spend?

_____

Distributive Property

KEY Concept

The Distributive Property states that to multiply a sum by a number, you can multiply each addend by the number outside the parentheses. You can multiply a difference by a number in a similar way.

Example

$$2(4 + 5) = (2 \times 4) + (2 \times 5)$$
$$2 \times 9 = 8 + 10$$
$$18 = 18$$

VOCABULARY

Distributive Property of Multiplication
  to multiply a *sum* by a number, you can multiply each *addend* by the number and add the *products*

Example 1

Use the Distributive Property and a model to find 5 × 12.

1. Draw a model to show 5 × 12.

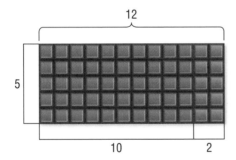

2. Draw a line to separate the factor 12 into tens and ones places.
   5 × 12 = 5 × (10 + 2)

3. Multiply to find the two products.
   5 × 10 = 50
   5 × 2 = 10

4. Add the products.
   50 + 10 = 60   So, 5 × 12 = 60.

YOUR TURN!

Use the Distributive Property and a model to find 3 × 13.

1. Draw a model to show 3 × 13.

2. Draw a line to separate the factor 13 into tens and ones places.

   _____

3. Multiply to find the two products.

   _____   _____

4. Add the products.

   _____

   _____

GO ON

Example 2

Use the Distributive Property to find 3(10 − 2).

1. Use the Distributive Property.

2. Simplify inside the parentheses.

3. Subtract.

$3(10 - 2) = (3 \times 10) - (3 \times 2)$

$= 30 - 6$

$= 24$

YOUR TURN!

Use the Distributive Property to find 5(7 + 1).

1. Use the Distributive Property. _____

2. Simplify inside the parentheses. _____

3. Add. _____

Who is Correct?

Use the Distributive Property to find 3(40 − 7).

Madela

$3(40 - 7) = 40 - (3 \times 7)$
$= 40 - 21$
$= 19$

Phil

$3(40 - 7) = (3 \times 40) - 7$
$= 120 - 7$
$= 113$

Juan

$3(40 - 7) = (3 \times 40) - (3 \times 7)$
$= 120 - 21$
$= 99$

Circle correct answer(s). Cross out incorrect answer(s).

▶ Guided Practice

I   Use the Distributive Property and a model to find 3×14.

Multiply to find the two products.

_____

Add the products.

_____

_____

2 Use the Distributive Property to find $9(7 + 4)$.

Step 1 Use the Distributive Property.

$9(7 + 4) = (9 \times \underline{\hspace{1cm}}) + (9 \times \underline{\hspace{1cm}})$

Step 2 Simplify inside the parentheses.

$(9 \times 7) + (9 \times 4) = \underline{\hspace{1cm}} + \underline{\hspace{1cm}}$

Step 3 Add. $63 + 36 = \underline{\hspace{1cm}}$

Use the Distributive Property to find each product. Show your work.

3 $6(6 - 4) = (\underline{\hspace{1cm}} \times 6) - (\underline{\hspace{1cm}} \times 4)$

$= \underline{\hspace{1cm}} - \underline{\hspace{1cm}}$

$= \underline{\hspace{1cm}}$

4 $7(9 + 5) = (\underline{\hspace{1cm}} \times 9) + (\underline{\hspace{1cm}} \times 5)$

$= \underline{\hspace{1cm}} + \underline{\hspace{1cm}}$

$= \underline{\hspace{1cm}}$

5 $8(5 + 6) = \underline{\hspace{4cm}}$

6 $4(8 - 2) = \underline{\hspace{4cm}}$

Step by Step **Problem-Solving Practice**

Problem-Solving Strategies

☑ Use a model.
☐ Look for a pattern.
☐ Guess and check.
☐ Act it out.
☐ Solve a simpler problem.

Solve.

7 **BASKETBALL** Enrico made five 3-point baskets, and Derek made six 3-point baskets in their last basketball game. How many points did they score in all?

Understand Read the problem. Write what you know.
Enrico made _____ 3-point baskets.
Derek made _____ 3-point baskets.

Plan Pick a strategy. One strategy is to use a model. Draw a model to represent the number of baskets the boys scored.

5 baskets   6 baskets

Solve Show the number of points scored.

3 points each

$\underline{\hspace{1cm}} \times 5 + \underline{\hspace{1cm}} \times 6$

$= \underline{\hspace{1cm}} + \underline{\hspace{1cm}}$

$= \underline{\hspace{1cm}}$

3×5 3×6

Enrico and Derek scored _____ points in all.

Check Does your answer make sense? How many *baskets* were made in all? _____

GO ON

8 **SCHOOL**   Mateo answered 12 five-point questions and 12 two-point questions correctly on his last history test. What was Mateo's test score? Check off each step.

_____ Understand: I circled key words.

_____ Plan: To solve the problem, I will _____.

_____ Solve: The answer is _____.

_____ Check: I checked my answer by _____.

9 **MONEY**   Ahmed earned $15 each week for 7 weeks. He spent $6 each week for 7 weeks. How much money does Ahmed have left?

_____

10 **BASKETBALL**   Penelope made 15 two-point baskets and 15 three-point baskets during this basketball season. How many points did Penelope score during the season?

_____

11 **Reflect**   What do you *distribute* when you use the Distributive Property?

_____

_____

 Skills, Concepts, and Problem Solving

Use the Distributive Property and a model to find each product.

12 $4 \times 16 = $ _____

13 $5 \times 15 = $ _____

Write the value that makes each equation true.

14 $6 \times 27 = (6 \times 20) + (6 \times$ _____$)$

15 $5(12 - 4) = (5 \times$ _____$) - (5 \times 4)$

16 $15 \times 48 = (15 \times$ _____$) + (15 \times 8)$

17 $28(8 - 5) = (28 \times$ _____$) - (28 \times 5)$

Use the Distributive Property to find each product. Show your work.

18 $14(5 + 3) =$ _____

  _____

19 $12(9 - 2) =$ _____

  _____

20 $5(11 - 4) =$ _____

  _____

21 $8(6 - 3) =$ _____

  _____

22 $2(5 + 18) =$ _____

  _____

23 $4(8 + 15) =$ _____

  _____

24 $18(2 + 3) =$ _____

25 $3(8 + 13) =$ _____

26 $11(7 - 2) =$ _____

27 $4(12 - 5) =$ _____

Use the Distributive Property to solve. Show your work.

28 **FOOTBALL**   The Mustangs scored 3 touchdowns (each worth 6 points) and 3 field goals (each worth 3 points). How many points did the football team score?

_____

29 **MONEY**   Marisa earned $18 each day for 8 days. She spent $3 on lunch each day for 8 days. How much money does Marisa have left?

_____

30 **PARTIES**   Your parents are paying for a party at your favorite restaurant. There will be 20 people at the party. Use the menu. How much will your parents spend if everyone orders the hamburger/fries special with a large soda?

_____

_____

Tony's Corner Cafe
Lunch Menu
Hamburger/Fries Special $6.00
Meatloaf Special $7.00
Small Soda $ .60
Large Soda $1.00

31 **TRIPS**   You are part of a group of 7 friends planning a trip to the art museum. Admission to the museum costs $6. A bus ticket to the museum costs $3. What is the total cost of the trip?

_____

_____

GO ON

Vocabulary Check   **Write the vocabulary word that completes the sentence.**

32 The _____ Property states that to multiply a sum by a number, you can multiply each addend by the same number and add the products.

33 **Writing in Math**   The Distributive Property can be used to write $5 \times (4 + 4) = (5 \times 4) + (5 \times 4)$. Can the Distributive Property also be used to write $5 \times (7 + 1) = (5 \times 7) + (5 \times 1)$? Explain why or why not.

_____

_____

 Spiral Review

Use the Commutative and Associative Properties to find each sum or product. (Lessons 4-1, p. 144 and 4-2, p. 151)

34 $12 + 37 + 18 =$ (_____ + _____) + _____

$=$ _____ + _____

$=$ _____

35 $8 \times 7 \times 5 =$ _____ $\times$ (_____ $\times$ _____)

$=$ _____ $\times$ _____

$=$ _____

SHOPPING   **Elan and Fala spent the day shopping on Saturday. Their receipts are shown below.** (Lessons 4-1, p. 144 and 4-2, p. 151)

36 The expression $(6 + 3) + 2$ represents the amount of money Elan and Fala spent at SNACK Bag. Use the Associative Property to write a second expression that represents the amount they spent on snacks.

_____

37 The equation $20 + 7 = 27$ represents the amount of money Elan and Fala spent at Sports 1. Use the Commutative Property to write a second equation that represents the amount they spent on clothing.

_____

Order of Operations

KEY Concept

You must follow the **order of operations** to evaluate mathematical expressions correctly.

Order of Operations	Symbol
1. Calculate operations inside parentheses.	**(parentheses)** **[brackets]**
2. Multiply and divide in order from left to right.	× ÷
3. Add and subtract in order from left to right.	+ −

VOCABULARY

order of operations
rules that tell what order to use when evaluating expressions
(1) Calculate operations inside parentheses.
(2) Multiply and divide in order from left to right.
(3) Add and subtract in order from left to right.

Sometimes parentheses are used to set a number apart from other operations.

Example 1

Find the value of $3 - 2 + 12 \div 4$.

Use the order of operations. There are no grouping symbols.

$$3 - 2 + 12 \div 4 = 3 - 2 + 3 \quad \text{Multiply and divide from left to right.}$$
$$= 1 + 3 \quad \text{Add and subtract from left to right.}$$
$$= 4$$

From left to right, subtraction comes first in this problem.

YOUR TURN!

Find the value of $7 - 5 + 3 \times 3$.

Use the order of operations. There are no grouping symbols.

$$7 - 5 + 3 \times 3 = 7 - 5 + \underline{\hspace{1cm}} \quad \text{Multiply and divide from left to right.}$$
$$= \underline{\hspace{1cm}} + \underline{\hspace{1cm}} \quad \text{Add and subtract from left to right.}$$
$$= \underline{\hspace{1cm}}$$

GO ON

Example 2

Find the value of $58 - (16 \div 4) \times 3$.

$58 - (16 \div 4) \times 3 = 58 - 4 \times 3$ Calculate operations inside parentheses.

$\qquad\qquad\qquad\quad = 58 - 12$ Multiply.

$\qquad\qquad\qquad\quad = 46$ Subtract.

YOUR TURN!

Find the value of $20 \div 5 + (3 + 2) \times 12 - 6$.

$20 \div 5 + (3 + 2) \times 12 - 6 = 20 \div 5 + \underline{} \times 12 - 6$ Calculate operations inside parentheses.

$\qquad\qquad\qquad\qquad\quad = \underline{} + \underline{} - 6$ Multiply and divide from left to right.

$\qquad\qquad\qquad\qquad\quad = \underline{} - 6$ Add.

$\qquad\qquad\qquad\qquad\quad = \underline{}$ Subtract.

Who is Correct?

Find the value of $10 \div 2 + (2 + 2) \times 2$.

Moria
$10 \div 2 + 2 + 4$
$= 5 + 2 + 4$
$= 7 + 4$
$= 11$

Corey
$10 \div 2 + (2 + 2) \times 2$
$= 5 + 4$
$= 9$

Anaba
$10 \div 2 + (2 + 2) \times 2$
$= 10 \div 2 + 4 \times 2$
$= 5 + 8$
$= 13$

Circle correct answer(s). Cross out incorrect answer(s).

 Guided Practice

Name the step that should be performed first.

1. $4 \times 6 + (30 - 3) \div 3$ _____

2. $26 \div 1 - (1 + 7) \times 2$ _____

3. $9 + 5 \div 5 + 2 \times 6$ _____

4. $8 + 2 \div 2 \times 5 - 1$ _____

5 Find the value of $70 - 6 \times (12 - 7) - 5 + 2$.

Step 1 Use the order of operations.
Calculate operations inside parentheses.
$70 - 6 \times (12 - 7) - 5 + 2 = 70 - 6 \times (\underline{\hspace{1cm}}) - 5 + 2$

Step 2 Multiply and divide.
$70 - 6 \times 5 - 5 + 2 = 70 - \underline{\hspace{1cm}} - 5 + 2$

Step 3 Add and subtract.
$70 - 30 - 5 + 2 = \underline{\hspace{1cm}} - 5 + 2$
$= \underline{\hspace{1cm}} + 2$
$= \underline{\hspace{1cm}}$

Find the value of each expression.

6 $9 - (1 + 8) + 3 = 9 - \underline{\hspace{1cm}} + 3$

$= \underline{\hspace{1cm}} + \underline{\hspace{1cm}}$

$= \underline{\hspace{1cm}}$

7 $8 \div 2 + (5 \times 2) - 7 = 8 \div 2 + \underline{\hspace{1cm}} - 7$

$= \underline{\hspace{1cm}} + \underline{\hspace{1cm}} - 7$

$= \underline{\hspace{1cm}} - 7$

$= \underline{\hspace{1cm}}$

8 $20 - 16 \div 4 \times 2 + (7 - 4) = \underline{\hspace{1cm}}$

9 $(4 \times 3) \times 5 \div 5 - (3 + 6) = \underline{\hspace{1cm}}$

10 $30 \div (2 + 3) + 2 \times 12 = \underline{\hspace{1cm}}$

11 $50 \div (7 + 3) \times 4 \div 2 = \underline{\hspace{1cm}}$

12 $(12 - 2) \div (4 + 1) - 0 = \underline{\hspace{1cm}}$

13 $15 - 3 + (2 \times 8) \div 4 = \underline{\hspace{1cm}}$

GO ON

Step by Step Problem-Solving Practice

Problem-Solving Strategies
- ☐ Draw a diagram.
- ☐ Guess and check.
- ☐ Act it out.
- ☑ Solve a simpler problem.
- ☐ Work backward.

14 NUTRITION Marcos has 2 baskets that hold 12 oranges each. He has 6 more baskets of 10 oranges each. Write and simplify an expression to find how many oranges Marcos has in all.

Understand Read the problem. Write what you know. There are baskets with _____ oranges each and baskets with _____ oranges each.

Plan Pick a strategy. One strategy is to solve a simpler problem. In this case, solving a simpler problem means to work on smaller parts of the expression, one at a time.

Solve Write an expression for the total number of oranges.

_____ × 12   +   _____ × 10
2 baskets of 12   6 baskets of 10

Simplify the expression using the order of operations.
$2 \times 12 + 6 \times 10$

= _____ + _____ Multiply.

= _____ Add.

Marcos has _____ oranges.

Check You can use addition to check.
$12 + 12 + 10 + 10 + 10 + 10 + 10 + 10 =$ _____

Write and simplify an expression to solve each problem.

15 NATURE Brandy likes to watch birds. She saw 1 wren make 2 nests and 3 wrens each make 4 nests. Five nests were damaged during a storm. How many nests were left? Check off each step.

_____ Understand: I circled key words.

_____ Plan: To solve the problem, I will _____.

_____ Solve: The answer is _____.

_____ Check: I checked my answer by _____.

16 COLLECTIONS   Tyler bought 3 packs of comic books. Each pack had 5 comic books. He gave 7 comic books to his brother. Then he bought 2 more packs of comic books with 18 books in each. How many comic books does Tyler have now?

_____

17 PHOTOGRAPHY   Marlene was looking through her photo album. She looked at 6 pages with 4 photos each. She removed 2 photos. Then she looked at 10 pages with 8 photos each. She removed 6 photos. Finally she looked at 2 pages with 2 photos each. How many photos were left in the album?

_____

_____

18 Reflect   Explain why $16 \div 2 + 6$ has a different value than $16 \div (2 + 6)$.

_____

_____

 ## Skills, Concepts, and Problem Solving

Name the step that should be performed first in each expression.

19 $3 \times 2 + 4 \div 2 - 2$ _____

20 $2 \times (3 - 6) + 9 \div 3$ _____

21 $(7 - 2 \times 4) - 8 \div 1$ _____

22 $4 + (6 - 2 \times 7) \div 2$ _____

23 $3(100 + 25) \times 2 - 25$ _____

24 $7 + 12 \div 15 \times 3$ _____

Find the value of each expression.

25 $50 \div 5 + 3 \times 2 - (8 - 2) =$ _____

26 $13 + 8 \div 2 - (10 + 2) =$ _____

27 $16 - 4 \times 0 + 18 - 15 =$ _____

28 $(9 - 6) + 8 \div 4 + 5 \times 6 =$ _____

29 $7 + 12 \times 3 - (4 - 4) =$ _____

30 $2 \times (14 - 6) \div 2 =$ _____

31 $8 \times (10 - 2) - 12 =$ _____

32 $21 + 4 \times 5 - (5 - 3) =$ _____

GO ON

Write and simplify an expression to solve each problem.

33 BOOKS   Ramona borrowed 2 stacks of books with 8 books each.
She then returned 9 books. Then Ramona borrowed 2 stacks of
5 books each. How many books does Ramona have now?

_____

34 COLLECTIONS   Don had 100 collector cards. He sold 5 packs of
baseball cards with 10 cards each. He then bought 3 packs of
football cards with 12 cards each. Then Don sold 25 hockey
cards. How many cards does Don have left?

_____

_____

Vocabulary Check   **Write the vocabulary word that completes each
sentence.**

35 Always calculate operations inside _____ first.

36 The _____ is a set of rules that tells what order to
follow when evaluating an expression.

37 Writing in Math   Does $30 - (10 - 5)$ equal $(30 - 10) - 5$? Explain.

_____

_____

 Spiral Review

Solve. Explain your reasoning. (Lesson 4-2, p. 151)

38 SPORTS   Amy bought the boxes of softballs
shown. Each softball cost \$4. How much did
Amy spend?

_____

Use the Commutative Property to fill in each blank.
Check your answer. (Lesson 4-1, p. 144)

39 $68 + 16 =$ _____ $+$ _____

_____ $=$ _____

40 _____ $+$ _____ $= 12 + 17$

_____ $=$ _____

STOP

Use the Distributive Property and a model to find each product.

1   $4 \times 13 =$ _____

2   $6 \times 16 =$ _____

Name each operation that should be performed first.

3   $8 - 4 \times (7 + 4) \div 2$ _____

4   $3 \times 2 - (12 \div 4) + 6$ _____

Use the Distributive Property to find each product. Show your work.

5   $7(3 + 2) =$ _____

6   $6(8 - 2) =$ _____

7   $3(9 + 5) =$ _____

8   $6(15 - 7) =$ _____

_____   _____

Find the value of each expression.

9   $18 + 12 \div 4 \times (5 - 2) + 7 =$ _____

10   $10 - (2 - 1) + 16 \div 2 \times (1 + 1) =$ _____

11   $28 \div 2 \times 8 + 4 \div 2 =$ _____

12   $64 \div 4 \times 5 - (30 - 18) \div 4 =$ _____

Solve. Explain your reasoning.

13   **BASKETBALL**   Joan made six 2-point field goals and two 3-point field goals. How many points did Joan score?

_____

14   **SHOPPING**   Payton had 50 pencils. He sold 3 bags of pencils with 5 pencils each. He then bought 2 packs of pencils with 10 pencils each. Then Payton gave 20 pencils to his sister. How many pencils does Payton have left?

_____

_____

Vocabulary and Concept Check

Associative Property of
Addition, *p. 151*

Associative Property of
Multiplication, *p. 151*

Commutative Property of
Addition, *p. 144*

Commutative Property of
Multiplication, *p. 144*

Distributive Property, *p. 159*

order of operations, *p. 165*

Write the vocabulary word that completes each sentence.

1 The property that states that the order in which two numbers are multiplied does not change the *product* is the _____.

2 The property that states that the grouping of the factors does not change the *product* is the

_____.

3 $(2 + 4) + 7 = 2 + (4 + 7)$ shows the

_____.

4 $3(2 + 5) = 3 \times 2 + 3 \times 5$ or $3(5 - 2) = 3 \times 5 - 3 \times 2$ is the

_____.

5 $3 + 9 = 9 + 3$ shows the

_____.

Write the correct vocabulary term in the blank.

6 _____

$$2(4 + 5) = (2 \times 4) + (2 \times 5)$$

7 _____

1) Calculate operations inside parentheses.

2) Multiply and divide from left to right.

3) Add and subtract from left to right.

Lesson Review

4-1 Commutative Property (pp. 144–150)

Use the Commutative Property to fill in each blank. Check your answer.

8 $2 \times 8 =$ _____ $\times 2$

$16 =$ _____

9 _____ $+ 7 = 7 + 15$

_____ $=$ _____

10 _____ $+ 4 =$ _____ $+$ _____

_____ $= 22$

Example 1

Draw a model to show $4 + 2 = 2 + 4$. Which property did you show?

1. Create a model for each side of the equation.

2. The order of the numbers changed, but the sum did not. This is the Commutative Property of Addition.

4-2 Associative Property (pp. 151–157)

Use the Associative Property to fill in each blank. Check your answer.

11 $8 \times (4 \times 3) = (8 \times$ _____ $) \times 3$

_____

_____

12 $(5 + 9) + 3 = 5 + ($ _____ $+ 3)$

_____

_____

13 $2 + (13 + 15) = ($ _____ $+$ _____ $) +$ _____

$2 +$ _____ $= 15 +$ _____

_____ $=$ _____

Example 2

Use the Associative Property to fill in each blank. Check your answer.

$(6 \times 3) \times 2 = 6 \times ($ _____ $\times 2)$

Use the Associative Property of Multiplication.
$(6 \times 3) \times 2 = 6 \times (3 \times 2)$

Check by multiplying the numbers on each side of the equation.
$(6 \times 3) \times 2 = 6 \times (3 \times 2)$
$18 \times 2 = 6 \times 6$
$36 = 36$

Draw a model to show each equation.

14 $(2 \times 3) \times 5 = 2 \times (3 \times 5)$
Which property did you show?

_____

_____

15 $(2 + 3) + 4 = 2 + (3 + 4)$
Which property did you show?

_____

Use the Commutative and Associative Properties to find each sum or product mentally.

16 $4 \times 7 \times 5 = 4 \times$ _____ $\times 7$

$\qquad = ($_____ $\times 5) \times 7$

$\qquad =$ _____ $\times 7$

$\qquad =$ _____

17 $11 + (9 + 4) = ($_____ $+ 9) + 4$

$\qquad =$ _____ $+ 4$

$\qquad =$ _____

Example 3

Draw a model to show $(1 \times 3) \times 4 = 1 \times (3 \times 4)$. Which property did you show?

1. Create a model for each side of the equation.

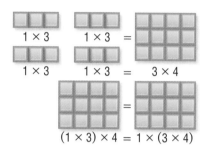

$(1 \times 3) \times 4 = 1 \times (3 \times 4)$

2. The grouping of the factors did not change the product.

Associative Property of Multiplication

Example 4

Use the Commutative and Associative Properties to find the sum mentally.

$15 + 9 + 5$

Determine which grouping would help you find the sum using mental math.
Group 15 and 5. Then find the sum.

$15 + 9 + 5 = 15 + 5 + 9$
$\qquad = (15 + 5) + 9$
$\qquad = 20 + 9$
$\qquad = 29$

4-3 Distributive Property (pp. 159–164)

Use the Distributive Property to find each product.

18 $9(5 - 3) = ($_____ $\times 5) - ($_____ $\times 3)$

$\qquad = $_____ $- $_____

$\qquad = $_____

19 $5(4 + 8) = ($_____ $\times 4) + ($_____ $\times 8)$

$\qquad = $_____ $+ $_____

$\qquad = $_____

Example 5

Use the Distributive Property to find $7(8 + 3)$.

$$7(8 + 3) = (7 \times 8) + (7 \times 3)$$

Solve within the parentheses. Then add.
$$(7 \times 8) + (7 \times 3) = 56 + 21$$
$$= 77$$

4-4 Order of Operations (pp. 165–170)

Find the value of each expression.

20 $10 + 8 \div 2 - 1 \times 3 = $_____

21 $11 - 7 + 3 \times 8 = $_____

22 $125 \div 5 \times 7 = $_____

23 $15 \div 3 + 4 \times 7 - 10 = $_____

Find the value of each expression.

24 $3 \times 4 - (4 - 2) + 6 \div 2 = $_____

25 $23 - 4 + (9 \times 2) \div 9 = $_____

26 $6 + 4 \div (8 - 6) \times 2 = $_____

27 $7 + 3 \times 5 + (6 \times 2) = $_____

Example 6

Find the value of $6 - 2 + 15 \div 3 \times 4$.

Use the order of operations.
There are no grouping symbols.
Multiply and divide.
$$6 - 2 + 15 \div 3 \times 4 = 6 - 2 + 5 \times 4$$
$$= 6 - 2 + 20$$

Add and subtract.
$$6 - 2 + 20 = 4 + 20$$
$$= 24$$

Example 7

Find the value of $18 \div 3 + (2 + 1) \times 4 - 5$.

Use the order of operations.
Calculate operations inside parentheses.
$$18 \div 3 + (2 + 1) \times 4 - 5 = 18 \div 3 + 3 \times 4 - 5$$

Multiply and divide.
$$18 \div 3 + 3 \times 4 - 5 = 6 + 12 - 5$$

Add and subtract.
$$6 + 12 - 5 = 18 - 5$$
$$= 13$$

**Use the Commutative Property to fill in each blank.
Check your answer.**

1   $4 \times 6 =$ _____ $\times 4$

    $24 =$ _____

2   _____ $+ 9 = 9 + 11$

    _____ $=$ _____

3   $7 + 9 =$ _____ $+$ _____

    _____ $=$ _____

4   $7 \times 12 =$ _____ $\times$ _____

    _____ $=$ _____

5   Give an example of the Commutative Property of Addition.
   Check your example.

_____

**Use the Associative Property to fill in each blank.
Check your answer.**

6   $3 \times (4 \times 10) = ($ _____ $\times 4) \times 10$

             $=$ _____ $\times 10$

             $=$ _____

7   $9 + (1 + 15) = ($ _____ $+ 1) + 15$

             $=$ _____ $+ 15$

             $=$ _____

8   $4 + (16 + 13) = ($ _____ $+$ _____ $) +$ _____

             $=$ _____ $+$ _____

             $=$ _____

9   $(7 \times 2) \times 5 =$ _____ $\times ($ _____ $\times$ _____ $)$

             $=$ _____ $\times$ _____

             $=$ _____

10   Give an example of the Associative Property of Multiplication.
    Check your answer.

_____

_____

**Use the Distributive Property to find each product.
Show your work.**

11   $9(2 + 5) =$ _____

12   $2(20 - 8) =$ _____

Solve.

13   $16 + 8 \div 4 \times 5 - (16 - 10) =$ _____

14   $(7 - 3) \times 3 \div 4 + (11 + 8) =$ _____

GO ON

Solve. Explain your reasoning.

15 COOKING Winston needs 3 gallons of stew for
a potluck dinner. Each gallon of stew requires
2 cans of beef gravy. Each can of beef gravy costs
$1.20. How much will the gravy for the stew cost?

gallon   gallon   gallon

_____

_____

16 SHOPPING Devin bought 4 boxes of markers with 5 markers in
each box. Katie purchased 5 boxes with 4 markers in each box.
Compare the number of markers.

_____

17 FITNESS Sofia participated in two 50-minute aerobics sessions
last week and three 50-minute aerobics sessions this week. How
many minutes did she work out during both weeks?

_____

18 POPULATION An apartment complex has 3 units. Four people
lived in each unit. Then 8 people moved away. The next month,
2 families of 5 moved into the complex. How many people live in
the apartment complex now?

_____

Correct the mistakes.

19 ART Terri purchased 3 boxes of paint tubes, each containing 8
tubes of paint. Raul purchased 8 boxes, each containing 3 paint
tubes. Raul told Terri that he bought more tubes of paint than she
did because he bought more boxes. What mistake did Raul make?

_____

_____

20 FOOD At the school cafeteria, the cook had 7 sandwiches. Two
students purchased 1 each. The cook then sold two teachers 2 each.
Her assistant made 11 more sandwiches. She dropped 1 of them on
the floor, so it had to be thrown away. The assistant said, "Now
you only have 10 sandwiches to sell." What mistake did she make?

_____

_____

Test Practice

Choose the best answer and fill in the corresponding circle on the sheet at right.

1 If $9 \times 8 \times 7 = 504$, then what is $7 \times 9 \times 8$?

    A 567   **C** 448

    B 504   **D** 343

---

2 $49 \times (130 \times 62) =$

    A $(48 \times 130) \times 26$

    B $(49 \times 130) \times 62$

    C $(49 \times 103) \times 62$

    D $(47 \times 130) \times 62$

---

3 Olivia wants to paint the two opposite walls in her bedroom. If she knows the dimensions of one wall, which expression can help her figure out the square footage of both walls?

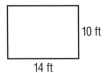

14 ft

    A $2(10 \times 14)$ **C** $2(14) \times 2(10)$

    B $2(14) + 10$ **D** $2 + (14 \times 10)$

---

4 $33 \div 3 \times 5 - (4 \times 3) =$

    A 954   **C** 43

    B 187   **D** 6

---

5 Which property is shown in the sentence below?

$$21 \times (8 + 5) = (21 \times 8) + (21 \times 5)$$

    A Associative Property of Addition

    B Distributive Property

    C Commutative Property of Addition

    D Identity Property of Multiplication

---

6 Which property is shown in the sentence below?

$$(346 \times 751) \times 203 = 346 \times (751 \times 203)$$

    A Associative Property of Multiplication

    B Distributive Property

    C Commutative Property of Multiplication

    D Associative Property of Addition

---

7 Which property is shown in the sentence below?

$$33 \times 1 = 1 \times 33$$

    A Associative Property of Addition

    B Distributive Property

    C Commutative Property of Multiplication

    D Associative Property of Multiplication

GO ON

8 Tyra has 24 minutes left in class. Before lunch, she has 2 more classes that are each 35 minutes and 1 that is 40 minutes. How many minutes does Tyra have before lunch?

A 94

B 114

C 134

D 84

9 What is the value of the number sentence?

$$25 \times (5 - 2) \div 5 - 12$$

A 285

B 363

C 3

D 15

10 $13 \times 52 \times 6 =$

A $52 \times 6 \times 15$

B $6 \times 13 \times 50$

C $13 \times 8 \times 52$

D $6 \times 52 \times 13$

11 Janet bought 5 packs of erasers with 10 erasers each. Jorge gave Janet 2 erasers. Then Janet gave 9 erasers to each of 3 friends. How many erasers does Janet have now?

A 15

B 30

C 25

D 47

12 Tamar sleeps 8 hours each night. Which expression represents how many hours he sleeps in a week?

A 8×7

B $8 + 7$

C $8 \div 7$

D $8 - 7$

ANSWER SHEET

Directions: Fill in the circle of each correct answer.

1 Ⓐ Ⓑ Ⓒ Ⓓ
2 Ⓐ Ⓑ Ⓒ Ⓓ
3 Ⓐ Ⓑ Ⓒ Ⓓ
4 Ⓐ Ⓑ Ⓒ Ⓓ
5 Ⓐ Ⓑ Ⓒ Ⓓ
6 Ⓐ Ⓑ Ⓒ Ⓓ
7 Ⓐ Ⓑ Ⓒ Ⓓ
8 Ⓐ Ⓑ Ⓒ Ⓓ
9 Ⓐ Ⓑ Ⓒ Ⓓ
10 Ⓐ Ⓑ Ⓒ Ⓓ
11 Ⓐ Ⓑ Ⓒ Ⓓ
12 Ⓐ Ⓑ Ⓒ Ⓓ

Success Strategy

If two answers seem correct, compare them for differences. Reread the problem to find the best answer between the two.

STOP

Index